THE EARTH SHELTER HANDBOOK

By Tri/Arch Associates: Gregory T. Baum, Andrew J. Boer, and James C. Macintosh Jr.

CONTENTS

ACKNOWLEDGEMENTS

We would like to give special thanks to the following people whom without their help this book could not have been written.

Professor Kent Keegan of the University of Wisconsin at Milwaukee School of Architecture.

The generous earth-sheltered home owners of our five county target area in northern Wisconsin, who let us tour their homes.

Allen Kunes of Oasis 2000, Rice Lake, Wisconsin for information and his hospitality.

Carl Rudenborg, Engineer, Menomonie, Wisconsin for design data.

Dorothy Gerasimo of Westcap, Menomonie, Wisconsin for information.

Christine Boer for proofreading and her patience.

Finally, Malcolm Wells for inspiration.

INTRODUCTION

This handbook was prepared through an independent research project at the University of Wisconsin at Milwaukee. Our mutual interest in earth-sheltered housing led us to combine our efforts to consolidate existing information necessary to begin siting, design, and construction of an earth-sheltered home. We define a home as earth-sheltered if earth is a major factor in the home's form and structure, resulting in site preservation and protection of the home from an adverse climate. Included are homes which use large earth berms, sod roofs, and those built below or into earth grades. After reviewing the existing information available on earth-sheltered dwellings, we felt a great need existed for more simplified and specific design data. This book is neither totally technical nor theoretically abstract but a mixture of both. Amateur energy enthusiasts as well as experienced professionals can relate to, and understand the information. Through several trips to northern Wisconsin, we encountered a lack of communication between people with unanswered questions about earth shelters and the people with those answers. This handbook was conceived as a device to combine, organize, and relay this information.

A target area in northern Wisconsin was selected including Barron, Chippewa, Dunn, Pierce, and St. Croix counties. The principles in this handbook, although intended to address the climate, soil types, and people of this target area, are valid for other regions.

Principles of site selection and preservation are presented by this handbook. Included also are graphic and verbal descriptions of various structural and building subsystems as well as construction details and techniques. In addition, applicable building codes, aesthetic values, and psychological effects pertaining to earth-sheltered homes are briefly discussed.

The book culminates in a matrix which summarizes the above information. This matrix simplifies the design process by matching different structural and building subsystems to specific sites. All data is coded through a numbering system which simplifies the use of the matrix.

The notion of building underground or using the earth for sheltering purposes is not new. Primitive insects to higher order mammals have developed very sophisticated underground building techniques. These were emulated by early man and are now beginning to be understood and appreciated today.

Earth shelters offer a viable alternative to the present above ground housing situation. Even though the current use of earth-sheltered housing is not widespread, we believe that their usefulness is far in excess of their current demand. In these times of inflation, with the cost of living increasing daily, earth-sheltered homes can reduce living expenses. They are a proven means of reducing energy consumption and lowering maintenance costs.

FRISCH, KARL *ANIMAL ARCH.*
HARCOURT, BRACE, JOVANOVICH, INC.
NEW YORK AND LONDON, 1974
 ● EXCERPTS TO RELATE TO
   EARTH SHELTERS

This spider burrows a tube-shaped sloping hole in a dry exposed bank, coats it with a mixture of soil particles and saliva, and thereafter lines it with silk threads, especially near the opening. A strong lid incorporating soil particles to give it weight is woven to cover the nest. This is fastened to the top edge of the opening by a strong silken hinge in such a manner that it can be easily opened but will close automatically by its own weight. The wide flange of this trap door is molded conically to ensure an exact fit to the mouth of the tube so that neither light nor rain water can get it. Its outer side is always camouflaged with soil from its immediate surroundings so that it blends with the environment and almost completely conceals the entrance to the tube.

The Trap-door Spider

1 PSYCHOLOGY

Minimal research on the psychological effects of living in earth-sheltered structures is available. However, feelings of claustrophobia, dampness, the fear of being buried alive, constant cold, and the fear of sewage backup have often been associated with living underground. The following paragraph presents an analogy of a common underground lifeform, ants.

Ants have shown us that many of the problems of underground living and the fears associated with these problems are unfounded. They have inspired solutions to several of man's functional problems related to living underground.

To strengthen the walls of their subterranean passages, the ants employ a mortar, derived from soil particles and secretion from their own bodies. This mortar acts much like our cement mortar we place between concrete blocks. The ants have also devised a means to eliminate dampness and mold. They carry damp nesting material from the subterranean chambers to the surface and deposit it on the outside of the nest where the material is allowed to dry, eliminating mold formation. This causes the original surface layer to move down and then the cycle begins again. The surface of the mound has a great number of holes which serve as entrances. All of these are not permanent. At night and on cold days the ants plug them with nesting material just as we close our windows to keep heat in. Mound building is an important method for obtaining solar heat. In the late and early hours of the day when the sun is low on the horizon, a domed nest intercepts more solar rays than a flat one. Catching the late day sun warms the mound

Karl Frisch, Animal Architecture (1974) p. 108

of earth for many hours after sunset because the earth functions as a heat sink.

Ants have essentially overcome the problems of dampness, mold, openness, closeness, and heat storage. Ants are one example in nature who use earth as the ultimate provider. Now we will deal with these problems as they relate to a higher life form, man.

I get claustrophobia . . . I need to get out to see the sunshine . . . That basement feeling . . .

Some people have a fear of being in enclosed, cramped, windowless spaces. Access to daylight helps eliminate the claustrophobic feelings. In earth sheltered design, correctly positioning skylights, glass panels, and courtyards will help cope with these feelings (see skylight section for complete daylighting principles). A central atrium, with working and living spaces wrapped around, offers natural light and outdoor rooms. The same principle can be applied to sloping sites through semi-enclosed courts and/or balconies.

- SKYLIGHTS

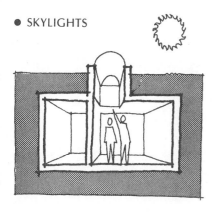

- SEMI-ENCLOSED COURTS
  - BALCONIES

- CENTRAL ATRIUM

I worry about being flooded out during a rainstorm . . . and the dampness . . .

● DRAIN TILE

An earth shelter has drain tile around its foundation and roof line to carry excess ground water away from the home. This reduces pressure on the walls and the waterproofing membrane.

Dampness is the result of dry concrete walls and interior air absorbing moisture from the ground. Various effective water and damp-proofing membranes which have been developed to impede moisture penetration are discussed later.

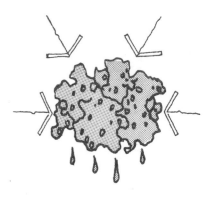

Yes . . . but isn't it cold down there . . .

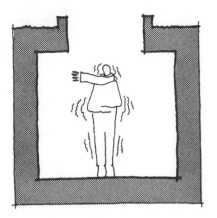

● WINTER, EARTH AND
SNOW INSULATE

● SUMMER, GROUND AIDS
IN COOLING

The soil surrounding an earth-sheltered home reduces the effect of severe climate temperature fluctuations. In the winter time at a depth of 10 feet the ground temperature is 45° F. The home is only required to be heated 25° to reach a comfortable living temperature. Insulations are available, which when applied correctly, significantly reduce heat transfer through walls, ceilings, and floors. Besides trapping heat in the winter time it prevents heat from entering in the summer. Due to its mass and the vegetative covering, the sod-covered roof reduces and delays heat gain from above.

I'm afraid sewage is going to back up and fill my
home . . .

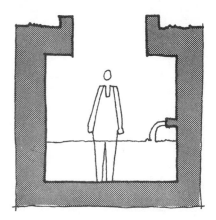

● TRADITIONAL WASTE
SYSTEMS

There are two approaches to sanitary waste
removal, a traditional one and an ecological one.
The traditional systems, cesspool, drainage field,
sand filter, and city hookup can be used with the
same results as an above grade home with or
without a pump (see waste disposal). The
ecological approach using waterless toilets or waste
digestors recycles organic waste generated in a
home. Shredded trash, paper, food scraps, and
treated sewage, can be distributed as a fertilizer on
a roof garden.

● ECOLOGICAL WASTE
SYSTEMS

I feel like the roof is going to cave in on me and all that earth will bury me alive . . .

The structural system of an earth-sheltered home is substantially stronger than its above grade counterpart. It must meet the higher demands of soil and snow loads. Strong and efficient structural systems minimize building failure from tornadoes, hailstorms, windstorms, and even aircraft accidents.

EARTH SHELTERS MINIMIZE
FAILURE FROM TORNADOES
AND WINDSTORMS

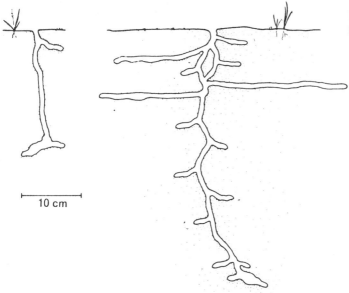

Karl Frisch, Animal Architecture (1974)
p. 106

10 cm

. . . in fact, the demands that a social community such as an ant colony makes on its home are considerably greater than those of a solitary insect. The various excavations must follow a definite plan and serve a variety of purposes. In a young nest, the lowest chamber is the sheltered place where the brood is housed and lovingly tended by the queen. When the first workers emerge, they extend the structure to greater depths and build horizontal galleries radiating from the main shaft in all directions. Normally, the deepest cavity continues to serve as the main breeding chamber, but in periods of rapid population growth some higherlying "residential chambers" may also have to be used to some extent for the brood. The passages closest to the surface are used for refuse.

Australian Wood Ant

2 ENVIRONMENTAL IMPACT

Topography

Soils

Sites

Orientation

Case Studies

Topography

Topography is the physical surface configuration of
land, including variations in surface elevation and
positions of natural and man-made features. The
earth's surface or subsurface, when rearranged or
disturbed, creates an imbalance in site stability,
often leading to problems with ecological systems,
accelerated runoff, and erosion.

BEFORE TREES ARE REMOVED

An ecological system is a community of plants and
animals cooperating and competing with each
other in the same environment, recycling nutrients,
and exploiting the sun's energy. Normally, and in
the absence of external changes, these systems
move toward a mature steady state in which they
produce their own food and dispose of their own
wastes. Forests, grasslands, and marshes depend on
specific temperature, moisture, and light levels to
remain in balance. When a foreign influence
interferes with one of these systems it has a far
reaching effect all along the food chain. For
example, when a series of trees are removed from
a wooded site, greater quantities of the sun's rays
reach the ground. This in turn, destroys ground
cover by drying out the soil, leaving it defenseless
against destructive erosion.

AFTER TREES ARE REMOVED

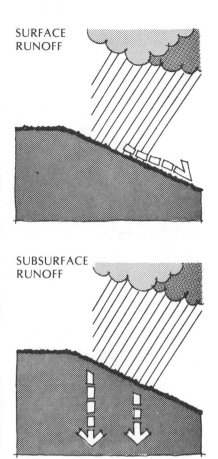

SURFACE RUNOFF

SUBSURFACE RUNOFF

RUNOFF

Surface runoff is that part of precipitation which flows over the ground. More importantly to earth shelters is subsurface runoff, which is the precipitation that infiltrates into and moves through the soil.

Surface Runoff

Before surface runoff can occur precipitation must be in excess of that required for evaporation, surface detention, and infiltration. When water vapor condenses and falls, a portion of that moisture evaporates in the air before it can reach the ground. Trees intercept precipitation at the treetop level and absorb it through their leaves. In addition, ground cover and organic matter detain and absorb a good portion of precipitation as well. The infiltration rate is dependent on vegetation, moisture content, and the size and arrangement of pore spaces in the soil. Sandy soils have large pore

EVAPORATION

SURFACE DETENTION

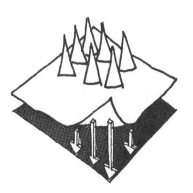

INFILTRATION

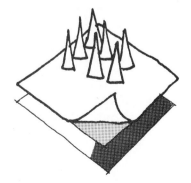

spaces, and a higher infiltration rate than silts and clays, which have relatively small pore spaces. Moisture content will be further discussed in the section on soils.

Subsurface Runoff

A more immediate concern pertaining to earth-shelters is subsurface water flow. Water penetrates the soil to three distinct levels: 1) water held near the surface by root systems, 2) water that penetrates to a semi-impervious layer, usually rock or clay, and moves laterally along the layer to surface drainageways, and 3) water moving directly to the water table.

The water table is the underground level below which all voids between soil particles are filled with water. Normally, this is a sloping flowing surface which follows the topography above and slopes down to ponds, lakes, and streams where it meets the ground surface. However, its depth below ground can vary tremendously and may fluctuate on a seasonal basis. Impervious subsurface material can also modify the water table, trapping water above or below. A high water table causes difficulties in excavation, as well as flooded basements, and unstable foundations. A high water table is indicated by springs, discolored darkened soils, the water levels in existing wells, and especially the presence of moisture loving vegetation such as willows, poplars, and reeds.

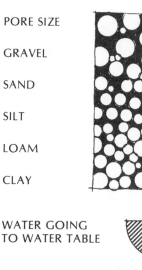

PORE SIZE

GRAVEL

SAND

SILT

LOAM

CLAY

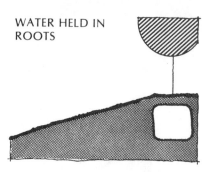

WATER GOING TO WATER TABLE

WATER HELD IN ROOTS

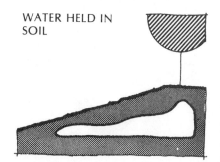

WATER HELD IN SOIL

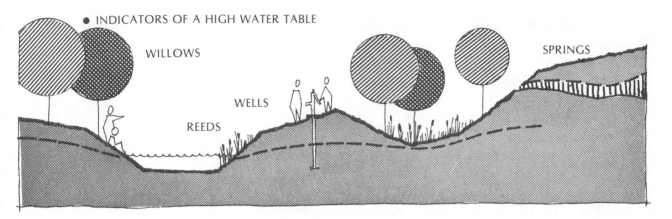

● INDICATORS OF A HIGH WATER TABLE

WILLOWS

SPRINGS

WELLS

REEDS

EROSION

When man interferes with and disrupts the ecological process it often results in an imbalance with extremely undesirable effects. One of the most noticeable is soil erosion.

Soil erosion is the wearing away or loss of soil by action of water and wind often caused by the lack of proper ground cover. The degree or severity of soil erosion is determined by soil character, degree of ground slope, and exposure to wind.

EXCAVATION WITHOUT
REPAIR LEADS TO
EROSION

Two serious effects of erosion are the reduction of the infiltration rate and water holding capacity of the soil. Soil profiles, to be discussed later, indicate that subsoils are lower in organic matter and not as permeable as topsoils. If the more absorbant topsoil is removed by excavation or erosion, the less permeable subsoil is exposed and will not absorb rainwater as rapidly. Consequently, there will be more runoff and less moisture retention.

Knowledge gained from the appearance of the ground surface can help predict subsurface conditions.

Soils

Soil profiles contain several layers of soil serving different functions. The organic layer, formed from rock and decaying plants, aids the soil's water holding capacity. The topsoil, a mixture of mineral and organic matter, is usually dark in color and holds plant nutrients. The substrata, largely mineral in composition and located below most plant root systems, functions as a sponge for waste material.

SOIL PROFILE

ORGANIC LAYER

TOPSOIL

SUBSTRATA

Charts, indicating type and location of soils, including descriptions, can be obtained from any state soil conservation office.

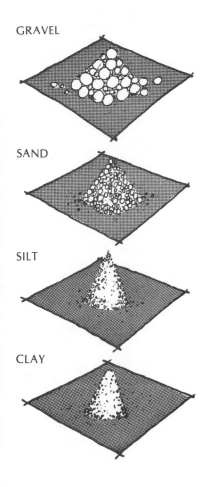

GRAVEL

SAND

SILT

CLAY

SOIL IDENTIFICATION

Soil particles are identified by their grain size. This is important because it allows soil to be broken down into one or more of the following classes. The chart below illustrates the major groups.

Gravel— particles over 2mm in diameter

Sand— gritty, finest particles visible to the naked eye, .05-2mm

Silt— invisible to the eye but can be felt, .002-.05

Clay— smooth and floury or clumpy when dry, plastic and sticky when wet, .002 and smaller

Organic—decomposed vegetation matter

It is possible to make a rough identification of a soil type in the field and analyze its properties with regard to bearing capacity (maximum allowable weight a soil can support), permeability, and water holding capacity. This rough identification is a useful process and in most cases may be all that is required to site a light structure. For this purpose soils are divided into mixtures of gravels, sands, silts, and clays which can be identified in the field.

Soil Classification

The clean—gravels	dominant material is gravel, less than 10% is silt or clay
Silty and clayey gravel—	mostly gravel, with more than 10% silt or clay
Clean sands—	dominant material is sand, less than 15% silt or clay
Silty and clayey sands—	dominant material is sand, with more than 10-12% silt or clay
Nonplastic—silts	very fine sands, liquid limit is less than 50% (those soils which begin to flow like a liquid when containing less than 50% water)
Plastic silts—	very fine sands, with a liquid limit over 50%
Organic silts—	silts dominated by organic matter, liquid limit under 50%
Nonplastic clays—	very fine clay, liquid limit under 50%
Plastic and organic clays—	liquid limit over 50% predominantly inorganic clay or a silt containing large amounts of organic matter

1

2

3

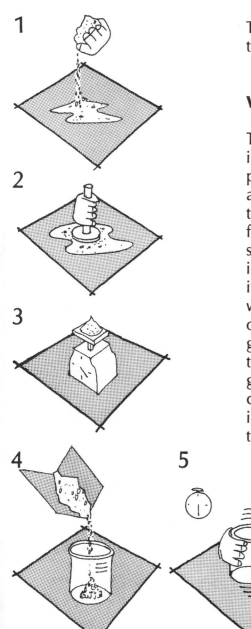

The ten soil classifications may be distinguished by the following field tests.

Visible Particle Size Test

Take a handful of soil as sample, dry it, and spread it out on clean paper. If more than half the particles are visible to the naked eye, it is a sand or a gravel. If this is difficult to determine even when the visible particles have been separated from the fine dust, then do the following: pulverize a dry sample, weigh it, cover it with five inches of water in a transparent container, shake it thoroughly, let it settle for thirty seconds, and then pour off the water. Continue to do this until the water poured off is clean. The remaining residue is the sand and gravel, and its dry weight can be compared with the original dry weight of the sample. In distinguishing a sand from a gravel, if more than half the coarse (visible) particles are over one quarter inch in size, it is a gravel, if not, it is a sand. If less than ten percent of the total soil sample is fine particles,

4   5   6   7

Kevin Lynch, Site Planning (1971)
p. 54

invisible to the eye or poured off in the sedimentation test, then it is a clean sand or a clean gravel. If not, it is a silty or clayey sand or gravel. A well graded clean sand or gravel contains all the particle sizes whereas a poorly graded one has significant gaps.

IF HALF THE PARTICLES ARE GREATER THAN 1/4" . . . IT'S A GRAVEL

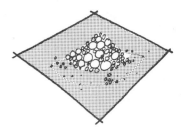

IF NOT GREATER THAN 1/4" . . . IT'S A SAND

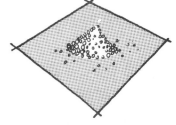

The following two tests must be used if the soil is not a sand or gravel. Both begin by picking out and discarding all the soil particles over about 1/64 inch in size. That is, the coarser ones that would interfere with molding and working the soil, and by taking a sample that will make a pat of soil about 1-1/2 by 1-1/2 by 1-1/2 inches thick.

IF 10% ARE INVISIBLE TO THE EYE . . . IT'S A SILTY SAND

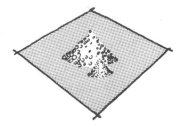

1

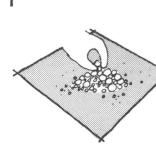

2

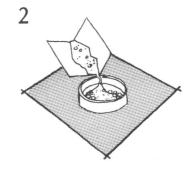

Kevin Lynch, Site Planning (1971) p. 54

1

2

Dry Strength Test

For the dry strength test, wet the soil and mold it into a pat. Take it between the thumbs and forefingers of both hands, and try to break it by pressure of the thumbs. If it cannot be broken or breaks only with great effort, snaps like a crisp cookie, and cannot be powdered then the soil is a plastic clay. If it can be broken or powdered with some effort, it is an organic clay or a nonplastic clay. If it is broken, powdered easily, or crumbles even as it is picked up, it is a plastic silt, an organic silt, or a nonplastic silt.

3   **4**   **5**

3   **4**   **5**

3   **4**   **5**

Kevin Lynch, Site Planning (1971)
p. 54

Thread Test

Add just enough water to a pat of soil so that it can be molded without sticking to the hands. On a nonabsorptive surface, such as a piece of glass, roll it out into a thread about 1/8 inch in diameter. Then remold it into a ball. If this process can be repeated without cracking, then the soil is a plastic clay. If the ball cracks, it is a nonplastic clay. If it cannot be remolded into a ball, it is a plastic silt or a plastic organic silt. If it cannot even be rolled into a thread, it is a nonplastic silt. In the course of this test the organic soils will feel spongy to the fingers.

1 WET THE SOIL

2 ROLL IT ON A PIECE OF GLASS

3 PICK UP THE ROLL

4 REMOLD IT INTO A BALL

5 IF IT CAN BE REMOLDED . . . IT'S A PLASTIC CLAY . . .

3 PICK UP THE ROLL

4 REMOLD IT INTO A BALL

5 IF IT CRACKS . . . IT'S A NONPLASTIC CLAY . . .

IF IT FALLS APART IT'S A PLASTIC SILT

Kevin Lynch, Site Planning (1971) p. 55

The dry strength and thread tests in part confirm each other and in part serve to separate the borderline cases, distinguishing a nonplastic from an organic clay, a plastic from an organic one, and a plastic from a nonplastic silt. The organic silts and clays not only feel spongy, but tend to be darker in color and have a musty odor if heated when wet. Peat or muck is easily identifiable. It is black or dark brown in color, has visible plant remains, a very spongy feel, and an immediate organic odor.

The above mentioned tests serve to distinguish the ten soil classes. Other field indicators exist such as the spongy feel of organic soils and the soapy feel of plastic clay. Pulverized clay feels smooth and floury. A dry lump sticks when lightly touched by the tongue. The organic soils are dark, drab grays, browns, and blacks.

Once a soil condition on a site is identified, the matrix on page 185 recommends preferable structural and waste disposal systems for the soil condition.

Kevin Lynch, Site Planning (1971)
p. 55

SOIL IMPLICATIONS

Some implications of these soils are summarized graphically below. The + , O , and − signs are used to indicate good, fair, or poor conditions, respectively.

	stability when loaded	stability when frozen	drain-age	erosion hazard
clean gravel	+	−	+	O
silty/clayey gravel	+	O	−	+
clean sand	+	+	+	+
silty-clayey sand	O	−	−	+
nonplastic silt	O	−	−	+
plastic silt	−	−	−	+
organic silt	−	−	−	O
nonplastic silt	O	−	−	+
plastic/organic clay	−	O	−	O
peat/muck	−	O	O	−

All earth shelters depend on soil for their ultimate support. Therefore, structural integrity depends on soil type and its bearing capacity. While few problems arise with regard to bearing capacity, poor soils often require specific foundation systems which directly affect building form. General bearing capacities of these soils are summarized below.

Material	Allowable bearing capacity in tons/ft^2
Sedimentary rock	◭ ◭ ◭ ◭ ◭ ◭ ◭ ◭ ◭
Dry clay	◭ ◭ ◭ ◭ ◭ ◭ ◭ ◭
Gravel; compact	◭ ◭ ◭ ◭ ◭ ◭
Sand; compact	◭ ◭ ◭ ◭ ◭
Sand; coarse & loose	◭ ◭ ◭ ◭
Sand; fine & loose	◭ ◭ ◭ ◭
Medium stiff clay	◭ ◭ ◭
Medium soft clay	◭ ◭ ◭
Soft clay	◭ ◭
Peat & organic soils	◭

It is important to remember that these are general values for each soil type and that they will vary from site to site. When in doubt about a soil or its bearing capacity, consult a qualified soils engineer.

Gravel is a well drained stable material and has a high bearing capacity when well graded.

Sand is well drained and makes a good foundation if well graded. Loose sands and gravels may settle under a load which in turn may cause it to lose its good internal drainage. Fine sands may flow like a liquid when saturated and uncontained.

Silt is stable when dry or damp, although it will compress under a load and be extremely unstable when wet. It also swells and heaves when frozen. As a result, foundations must be flexible or of sufficient depth to prevent this action. The erosion of silt is likely to be severe.

Clay is very loose and pliable when wet, stiff and cohesive when dry. Its reaction to frost is less severe than silty soil. It tends to resist the passage of water, slip, swell, or soften when wet. It can be a good bearing soil when kept dry.

Primarily peat, and to a lesser extent other organic soils are poor load bearing soils. Peat is spongy, weak, and has little cohesion. Cohesion is the ability of the soil to adhere to itself. Under normal circumstances peat is removed from a site or left undeveloped.

Clean sands and gravels are good for sewage drain fields. All other soil types must be checked for their absorptive capacity before use. It is likely that other sands and gravels as well as inorganic silts would be acceptable. Clays and organic soils should be avoided.

Sites

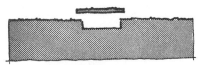

NEARLY FLAT SITE

Analysis of slope conditions will include: nearly level sites, 0-6%; moderately sloping sites, 6-10% steep slopes with a building at the top, 10-30%; and steep slopes with a building at the bottom, 10-30%.

The degree of slope also has connotations pertaining to the type of soil the slope contains. Gravels, sands, and silts were predominant in our analysis of Barron County, Wisconsin. Clays were not present in our target area, thus being omitted.

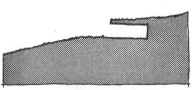

MODERATELY SLOPING SITE

NEARLY FLAT SITE

Two basic soil types were found for a 0-6% slope in Barron County. They were Arland fine sandy loam and Barronett silty loam.

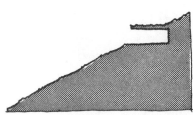

STEEP SLOPE, TOP SITE

Arland Fine Sandy Loam

Arland fine sandy loam is characterized by a grayish-brown color, content of little organic material, a gravel and stone mixture, and a strong acid content. The advantages of building on a fine sand far surpass the disadvantages. Accelerated runoff and erosion are virtually nonexistent because of its rapid surface and subsurface drainage. Another drainage advantage of sandy loam is its ability to accept waste material.

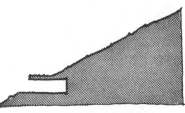

STEEP SLOPE, BOTTOM SITE

Generally, fine sands are good foundation material. Arland sandy loam contains gravel and some stone which helps the sand keep its good drainage characteristics when loaded, thus reducing excess foundation wall pressure. Arland may be a loose sand, that is, not dense, close, or compact in structure. It may settle initially under a load. This is not a serious problem if the soil is consistent under the entire structure which allows it to settle uniformly.

Barronett Silt Loam

Barronett silt loam is characterized by a dark grey color, a crumbling nature, a thin platy to soft crumb structure and a strong acid content. This soil, being a silty loam, has many inherent disadvantages. One disadvantage is ponding of water. Ponding is primarily caused by the soil's poor drainage characteristics. Vegetation can help reduce this problem. Willows and poplars should be planted where ponding occurs to soak up excess moisture. Ponding may become more critical when the surrounding terrain drains toward the site. This condition requires extensive drain tile usage and berming. A swale may be constructed to carry away off-site water. The swale should be placed as far away from the structure as possible. One method to reduce excess pressure from the walls and foundation system uses burlap bags filled with gravel set vertically along the foundation wall.

CONSISTENT SAND WILL SETTLE UNIFORMLY . . .

ON-SITE SPOT PONDING

WILLOWS WILL SOAK UP EXCESS MOISTURE . . .

BERM/SWALE WATER FROM OFF SITE TERRAIN . . .

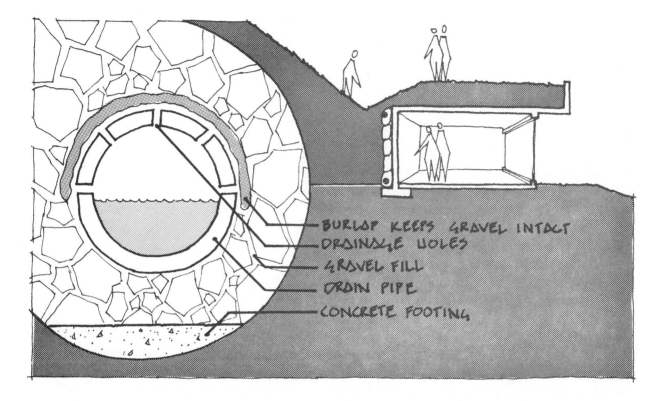

BURLAP KEEPS GRAVEL INTACT
DRAINAGE HOLES
GRAVEL FILL
DRAIN PIPE
CONCRETE FOOTING

Another disadvantage of a silty loam is that it must pass a test for absorption before a waste system can be selected.

One last disadvantage of Barronett silt loam is that it is extremely unstable when wet. It also swells and heaves when frozen, thereby calling for special foundation systems. Standard footing sizes become inadequate because of the soil's low bearing capacity. The use of larger footings, approved and/or designed by an engineer, is recommended. It is also important to remember that the key to a successful foundation is uniform settlement.

MODERATELY SLOPING

Two basic soil types exist for a 6-10% slope in Barron County. They are Santiago silt loam and Hixton fine sandy loam.

Santiago Silt Loam

Santiago silt loam is characterized by a greyish-brown color, little organic matter, and a strong acid content. Santiago silt loam drains very slowly thus creating rapid runoff. This leaves the potential for erosion extremely high. The rate of runoff can be reduced significantly by efforts to decrease the velocity of rain and to increase the absorptive capacity of the soil.

The velocity of rain can be decreased by leaves and branches of tree tops which hold raindrops momentarily. Existing trees on the site should be preserved and new trees introduced to take advantage of this effect.

TREE TOPS REDUCE THE
VELOCITY OF RAIN . . .

RAIN WATER CAN
FILL SPACES LEFT
BY DEAD PLANTS . . .

Vegetation is also the best means to increase the soil's absorptive capacity. Shrubs with fibrous roots should be planted to hold the permeable surface layer of soil in place. The roots will introduce air into the surface soil, thus creating larger pore spaces that can fill with rain. Grasses should also be introduced in a generous manner, not less than a 70% density, to hold the soil in place and absorb surface precipitation.

Due to Santiago silt loam's drainage characteristics, slope drainage fields and other disposal systems may be not feasible. Before a waste system can be installed, the absorptive capacity of the soil must be tested.

Hixton Fine Sandy Loam

Hixton fine sandy loam is characterized by a greyish-brown to yellowish-red color, thin flaky layers, and a strong acid content. Hixton fine sandy loam drains fairly rapidly but holds a great deal of water. This implies that accelerated runoff and/or erosion is not critical but subsurface water levels must be checked. The foundation system may disturb subsurface water because the water table

MODERATE SLOPES MAY HAVE HIGH WATER TABLES . . . IF DISTURBED THEY MUST BE REROUTED

WATER TABLE

may be relatively close to the surface on a moderate slope. When the flow of a water table has been interrupted by a wall of a building, a dam effect is created which produces excessive pressure buildup. This often leads to cracking which allows moisture penetration. The best method of protecting against water pressure build-up is to provide proper drainage of water away from the walls and foundations. A simple porous channel, parallel to the natural drainage path, will carry away the dammed water. In extreme cases, where hardpan or dense silty loams might prevent infiltration, gravel placed in burlap bags could be set vertically along the foundation wall down to the drain tile.

If earth is sloping toward the structure, construct a swale, gravel bed, or porous pipe, directing water around and away from walls and foundation. A swale is the easiest and least expensive technique and can be applied when the soil's drainage characteristics are fair to good. The swale should be densely planted with grass to prevent erosion. A gravel trench is most effective when it is used with a poor draining soil. It involves digging a trench and filling it with gravel or sand. If the trench must handle a great deal of water a drain tile may be added to handle greater amounts of water in a short period of time.

In most cases a sandy loam is suitable for sanitary waste disposal, but it should be tested for its absorptive capacity to determine which waste system should be used.

● SWALE

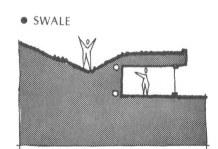

● GRAVEL TRENCH

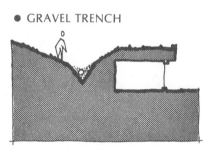

● SLOPE AWAY W/DRAIN

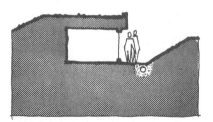

DRAINAGE
TECHNIQUE . . . SIMPLE
TERRACE . . .

SAND
WOOD
PITCH

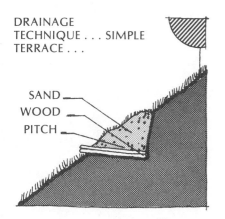

ON STEEP SLOPES 30%
OR MORE . . . A ROW
OF TERRACES MAY
BE NECESSARY . . .

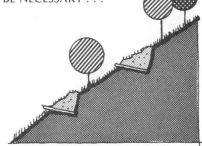

LONG TERRACES CAN BE
PITCHED TO DEPOSIT ALL
EXCESS WATER AROUND
SITE . . .

STEEP SLOPE, TOP

Two basic soil types were found for a 10-30% slope in Barron County. They include Antigo silt loam and Omega loam sand.

Antigo Silt Loam

Antigo silt loam is characterized by a brownish-gray to yellowish-brown color, thin flaky layers, and a strong acid content.

This soil has poor drainage characteristics and bearing capacity. The previously mentioned techniques to control runoff are applicable but may not suffice. A more rigorous technique such as terracing is needed. A terrace can be simply constructed using sand and wood. A block of soil is removed and replaced by a layer of wood with sand piled on top. This will allow accelerated runoff to infiltrate into the soil, or if pitched correctly, will redirect water around a structure. If the slope is in excess of 30% several terraces may be necessary to prevent erosion. Antigo silt loam holds large quantities of water, consequently it may remain saturated quite frequently. The absorptive capacity of Antigo silt loam must be tested to determine which waste disposal system should be employed.

The bearing capacity of a silty loam must be analyzed when buildings are near the top of a slope. When the weight of a building exceeds the soil's bearing capacity, slump may occur. To prevent this action a lighter structural system should be selected, deeper foundations should be dug, or an alternative site should be considered.

Omega Loamy Sand

Omega loamy sand is characterized by a very dark brown to brown color and a strong acid content. This soil has few of the problems of Antigo silt loam.

Any unrepaired excavation will encourage erosion from wind. When wind sweeps over exposed dry soil the smaller lighter particles are lifted and held in suspension as dust. Slightly heavier particles are also suspended when the wind velocity is greater. However, the larger particles, too heavy to be carried by the wind, drop back down to the surface, loosening other particles. The scouring process, called saltation, causes smaller particles to become dislodged and windborne every time a larger particle falls, thus accelerating soil erosion.

An effective way to cope with wind erosion is through the planting of dense vegetation. Ground cover of 70% density is the most successful means to hold soil in place. It not only reduces wind velocity but also holds the soil together with its fibrous roots. Some ground covers may not thrive in dry soil so selection should be carefully made. If ground cover, such as grass, will not grow well, then coniferous shrubs used as windbreaks can be substituted to retard erosion by reducing the wind velocity. Local wind patterns must be obtained to determine where a wind break will be most effective.

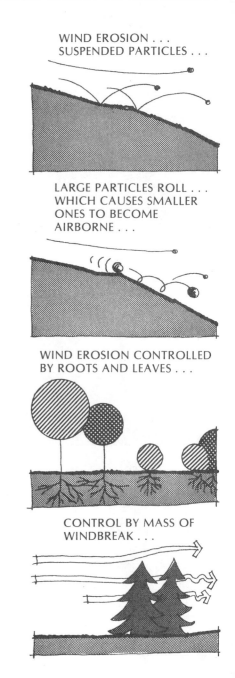

WIND EROSION . . .
SUSPENDED PARTICLES . . .

LARGE PARTICLES ROLL . . .
WHICH CAUSES SMALLER
ONES TO BECOME
AIRBORNE . . .

WIND EROSION CONTROLLED
BY ROOTS AND LEAVES . . .

CONTROL BY MASS OF
WINDBREAK . . .

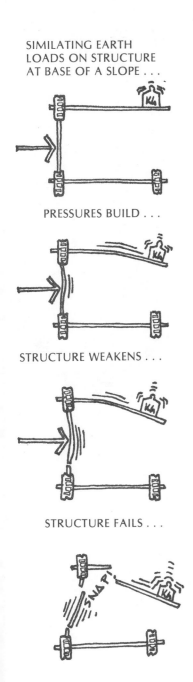

SIMILATING EARTH
LOADS ON STRUCTURE
AT BASE OF A SLOPE . . .

PRESSURES BUILD . . .

STRUCTURE WEAKENS . . .

STRUCTURE FAILS . . .

Omega loamy sand is a good load bearing soil. However, when building on a steep slope it is important to place the footings into previously undisturbed soil.

STEEP SLOPE, BOTTOM

The same two basic soils found on a steep slope with development at or near the top, were also found on a steep slope with a building placed at the bottom. They are Antigo silt loam, and Omega loamy sand.

Antigo Silt Loam

Antigo silt loam at the base of a steep slope may be the worst condition for development. The soil has weak bearing capacity when loaded and poor drainage characteristics.

Antigo silt loam may slump when disturbed at the bottom of a steep slope. The removal of a large block of soil at the base of a steep slope, to allow for an earth shelter, must be replaced with a structural system equal to or greater than the lateral force resistance of the block removed. In addition to earth forces, resisting water pressure must be included in the design of the structural system.

The location of the water table becomes critical when locating an earth shelter at the base of a steep slope. Building below the water table is not advisable because the water cannot be diverted. Before building on the site, locate the water table by drilling on the slope or by talking to neighbors. It is advisable to consult a soils engineer for drainage information and a structural engineer for design of the structural system.

LOCATING THE WATER TABLE . . . TEST DRILLS . . . . .

TALK TO NEIGHBORS . . .

Omega Loamy Sand

Omega loamy sand is a fairly good soil for development at the bottom of a steep slope. Its structural and drainage characteristics are favorable. Omega loamy sand is generally well packed and should provide a consistent base for a foundation system. Again, the structural system must be equal to or greater than the block of soil removed. Backfilling with gravel is recommended even when the soil's drainage characteristics are favorable. Larger drain tile is recommended to handle the larger volume of water moving down the slope. Drain tile should be placed at the top as well as at the base of the structure to reduce water pressure.

Orientation

One major aspect of site planning is the orientation of the structure. Although the structure is earth sheltered, it should not be thought of as being completely buried. Necessary doors and windows are usually exposed. The grouping and direction of these openings is referred to as the orientation of the structure. The major design elements are sun and wind. Proper orientation with respect to these elements can provide significant energy savings.

SUN

The sun is the source of solar radiation as well as natural light. To be energy efficient a home must take advantage of these basic properties of the sun. When building above the 40° North Latitude the first concern is collecting the winter sun and the second is shading summer sun.

The position of the sun relative to the building is an important factor in designing for solar heat gain. With readings of the sun's azimuth and altitude, the position of the sun in relation to the time of day and season can be determined for the site.

COLLECT WINTER SUN . . .

BLOCK SUMMER SUN . . .

The azimuth is the angle of the sun's image superimposed upon the earth's surface. This is helpful when orienting a structure for winter solar gain as well as room layout. When designing a room layout, a decision must be made about which rooms should receive morning, afternoon, and evening sun.

SUN'S AZIMUTH

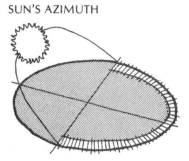

DESIGN TOOL . . .

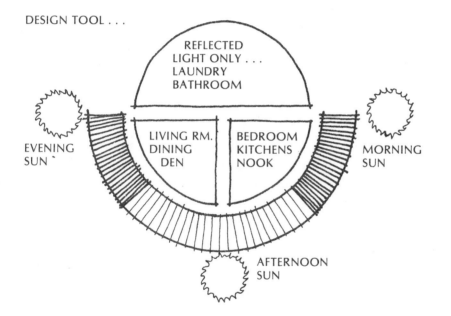

The sun's altitude is its angular elevation above the horizon. The altitude will be used to determine the size and location of shading devices on the building, as well as, the type and location of vegetation to be used for shading purposes.

Shading devices shield interior and exterior spaces of an earth shelter from solar radiation. Their effectiveness depends on their location and form. On the following page some basic types of shading devices are illustrated.

SUN'S ALTITUDE

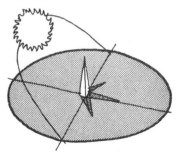

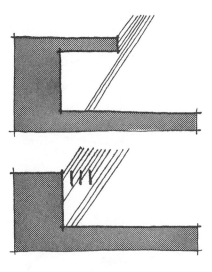

Overhangs

Horizontal overhangs are most effective when shading southern facades. They offer view and protection from the rain.

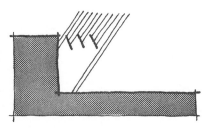

Horizontal Louvers

Horizontal louvers parallel to the wall permit air circulation near the wall to carry away hot stagnant air. They are most efficient in shading southern exposures.

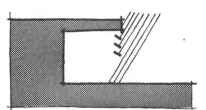

Slanted Louvers

Slanted louvers, in comparison to horizontal louvers, permit better protection from the sun yet maintain air circulation. The angle of louvers can be adjusted to the angle of the sun.

Louvers With Overhangs

Louvers with overhangs provide protection from low sun angles, typical of extreme eastern and western sun positions. The louvers will interfere with view.

Vertical Louvers

Vertical fins are most effective for eastern and western exposures. They too, can be adjusted to the angle of the sun, and will permit air circulation when separated from the wall.

Depending upon the relative shape and facade direction, as well as upon other factors, such as wind, the optimum orientation will vary for different sites. In latitudes above 40° North, south facades receive twice as many of the sun's rays in winter than summer. East and west facades receive two and one half times as many rays in summer than in winter. All structures elongated along the north-south axis are less efficient than a square structure during all seasons. The optimum form should be elongated in the east-west axis. The least efficient form is a square oriented NNE or ENE.

The quantity of the sun's rays which strike various structures of different form, size, and orientation will prove to be a useful design tool. The illustrated chart will help determine the placement of windows with respect to energy. The ratios in this chart are for January 21, 40° North Latitude.

The relative solar data implies that a structure elongated east-west has the greatest potential for solar heat gain. This is significantly greater than a structure elongated north-south. The addition of a second floor to an earth shelter will double the amount of total solar heat gain.

BEST ORIENTATION IS ALONG THE E-W AXIS

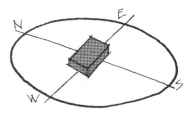

POOREST ORIENTATION IS ALONG THE ENE AXIS

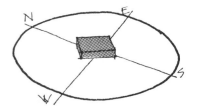

SQUARE

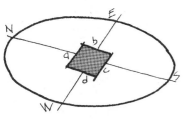

Percent of solar gain striking the facade when the home is located on a N-S, E-W axis.

N-S RECTANGLE

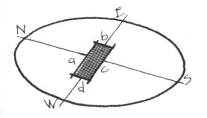

E-W RECTANGLE

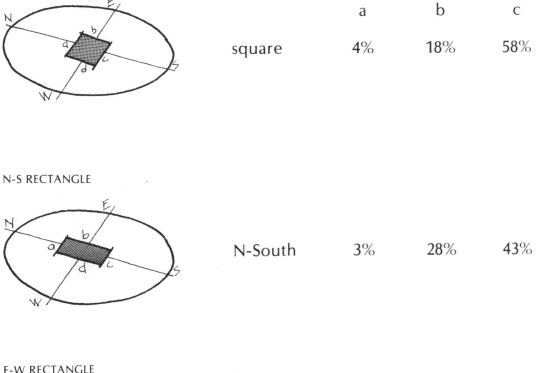

	a	b	c	d
square	4%	18%	58%	18%
N-South	3%	28%	43%	28%
E-West	5%	11%	72%	11%

Percent of solar gain striking the facade when the home is tilted 45° from the N-S axis.

	a	b	c	d
square	6%	56%	56%	6%
N-South	3%	60%	30%	6%
E-West	6%	30%	60%	3%

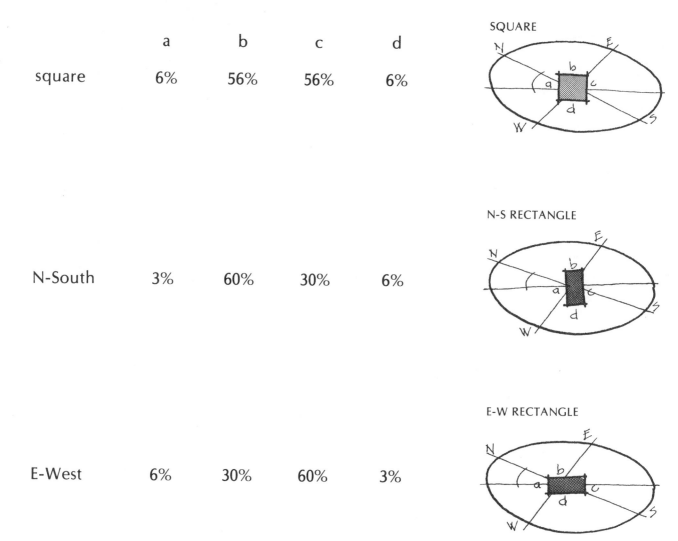

● SUMMER
BREEZES
FOR
COOLING

● NEED PROTECTION FROM
WINTER
WINDS

● WIND OVER THE
STRUCTURE . . .

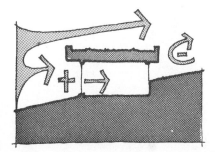

WIND

In seasonal climatic cycles, large volumes of wind move above the surface of the earth at varying speeds, intensities, and temperatures. These winds, which can have positive effects during the summer and negative effects during the winter, must be taken into account if the structure is to be energy efficient.

Winter Infiltration

During winter it is essential to minimize air infiltration into an earth shelter. The two main causes of air infiltration are wind movement, which creates positive and negative pressure areas outside the house, and secondly, negative pressure inside the home caused by various exhaust mechanisms. Of these, the first is most crucial.

The first cause of air infiltration is wind movement drawing air through the structure. On the windward side of the home a positive pressure builds up since the air is being compressed. On the leeward side a negative pressure is produced due to the suction of air away from that side. This pressure difference creates air infiltration into the structure through doors, windows, and cracks.

The second cause of air infiltration occurs when a negative pressure exists within the structure. This condition is created by a kitchen range hood or by venting a combustion unit. The oil burner or wood stove behaves in this manner; it takes air from an enclosed room, heats it, and vents it up the chimney. This results in a negative indoor pressure which draws in outside air, again through doors, windows, and cracks.

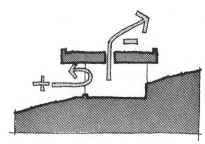

● AIR LOST THROUGH EXHAUST FANS . . .

Normal prevailing winds for the target area are from the northwest in winter. Local topography and vegetation may alter the prevailing wind patterns significantly. They must be studied carefully prior to orienting a home.

Summer Cooling

Summer cooling by natural ventilation is a prime consideration during site planning and preliminary design. Natural cooling, and more importantly, air circulation for summer, is often overlooked. To obtain maximum natural ventilation and cooling, it is important that there be an awareness of the wind velocity and direction, the location and size of vents, and humidity and temperature. Normal prevailing winds for the target area occur from the south in summer. However, wind patterns vary from site to site depending on local land forms and vegetation. The exit opening should be at least as large as the inlet opening to maximize air flow. The diagrams below illustrate both good and bad venting.

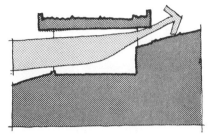

An undesirable situation exists when the inlet is larger than the outlet. In this condition the greatest amount of cooling takes place outside the structure.

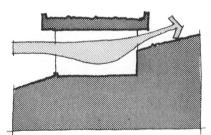

When the inlet and outlet are equally sized, good cooling results.

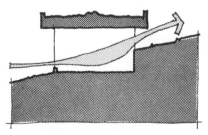

When humidity is high, the internal air velocity should be as great as possible because wind feels cooler when it's blowing faster. This is accomplished by decreasing the inlet opening one half the size of the outlet opening.

The height of the vent depends upon what part of the room is to be ventilated. For example, if the ceiling is to be vented, which is generally the case, air shall enter high and exit high.

All immediate obstructions hindering the passage of air to the vents should be moved. This includes any trees or other foliage which hinders air movement to the vents. This can also be corrected by placing the vents above, below, or to either side of a tree or interfering foliage.

CEILING VENT

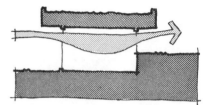

WINDOW VENT

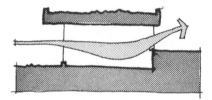

FLOOR VENT

Since the temperature decreases at night it may be advantageous to use night time breezes to carry away warm, stagnant, humid air which has accumulated during the day.

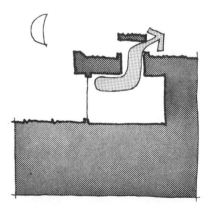

ROOF WARMS UP
DURING THE DAY . . .

EARTH ROOF HEATS UP
DURING THE AFTERNOON . . .

HEAT TRANSFERRED INSIDE
NEEDS AN ESCAPE VENT

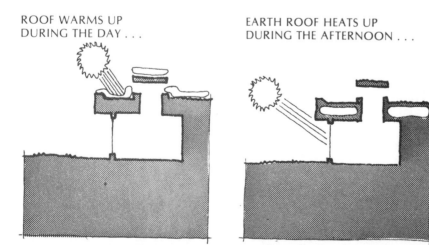

Case Studies

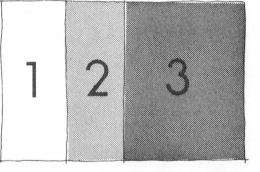

SUN

Steeply sloping, moderately sloping, and slightly sloping sites, in addition to orientation have been analyzed with respect to sun. The matrix below will help organize and identify specific problems.

	NORTH	SOUTH	EAST	WEST
FLAT				
MODERATELY SLOPING	1	2	3	
STEEPLY SLOPING, TOP				
STEEPLY, SLOPING, BOTTOM				

NORTHERN FACADE . . .

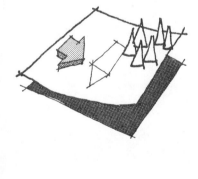

EXPOSE EAST/WEST FACADE FOR DIRECT SUNLIGHT

1 Northern Facades

These implications are relevant for slight, moderate, and steeply sloping sites. North facing facades will only receive reflected sun rays. To receive direct sun rays the addition of skylights, monitors, and clearstories are required. These windows should be positioned anywhere from SSE through SSW depending upon what time of day the direct rays are desired. These devices cannot capture early morning and late afternoon direct sun. The early and late sun can be captured by digging out a portion of the east or west facade and inserting a window panel. Good window ground clearance is always desirable.

2 Southern Facade

A facade oriented 17.5° east of south will maximize winter solar gain, as well as provide the longest period of sun in any structure. Overhangs, slanted louvers, horizontal louvers, or deciduous foliage must be used to block the intense summer sun.

3 East/West Facades

Any deviation from the north-south axis will change the normally equal distribution of the sun's rays. The total distribution between east and west rays is interchangeable. East facades receive morning rays, therefore bedrooms, breakfast nooks, and living rooms positioned along this facade will receive early morning and afternoon light. The usual late afternoon and evening glare has been avoided, and maintains an unobstructed view to the outdoors. If late afternoon sun is desired, incorporating a west facing skylight will capture and pleasantly diffuse the sun's rays throughout the room. West facing facades have the advantage of direct sun rays until sunset, but the disadvantage of late afternoon glare. Correctly positioning overhangs with louvers or vertical fins can reduce glare. Glare can result from reflections of the sun from lakes, asphalt surfaces, and windows of nearby homes. These devices restrict view depending upon how low the louvers hang or at what angle the fins are positioned. We recommend moveable fins and louvers so views are possible when glare is not a problem.

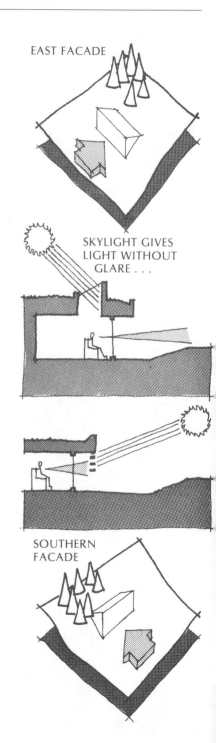

EAST FACADE

SKYLIGHT GIVES LIGHT WITHOUT GLARE . . .

SOUTHERN FACADE

SUMMER COOLING

Steeply sloping, moderately sloping, and slightly sloping sites, in addition to orientation have been analyzed with respect to summer cooling. The matrix below will help organize and identify specific problems.

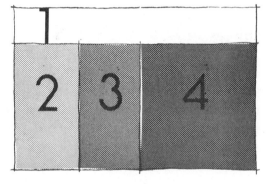

NORTH   SOUTH   EAST   WEST

FLAT

MODERATELY SLOPING

STEEPLY SLOPING, TOP

STEEPLY SLOPING, BOTTOM

FLAT SITE/ALL EXPOSURES

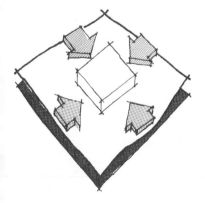

1 Flat Sites, All Exposures

All exposures on a flat site are able to capture southerly breezes for natural ventilation. These breezes can be captured by operable skylights, vents, and raised roofs. A north facing facade is being used although the information is applicable to all exposures.

A skylight, as shown, will both illuminate and ventilate. The skylight will supply natural air motion to allow musty stagnant air to escape. The air enters high through the vent carrying stagnant air out through another high vent positioned on the north facade. Low vents on the north facade will also aid in ventilation due to the negative

pressure created by the wind passing over
the structure. This skylighting design is an untested
conceptual solution which appears to be feasible.
However, it must be mentioned that because it is
not a standard skylight, escalated construction and
material costs are possible.

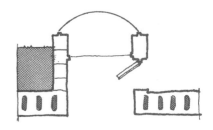

Another good venting technique is raising a
portion of the roof structure or installing a
monitor. These devices rely on the principle that
hot air rises and cool air falls. This principle
suggests that a negative pressure exists in the
structure which will draw cool air from the
outdoors in through a low positioned vent. Warm
stagnant air rises and escapes through vents in the
raised roof. Besides ventilating the structure, this
technique allows direct and diffused light into
many otherwise artificially lit rooms.

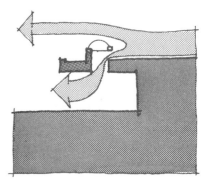

HOT AIR RISES AND IS SWEPT AWAY BY SUMMER BREEZES . . .

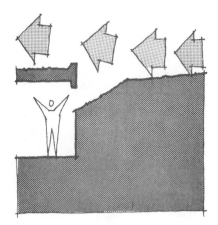

To vent on the south facade extensive earth moving is required. It is important to leave sufficient exposure in front of the vent when designing a berm. This means that the grade leading to the vent should be gradual to insure proper air intake. If the grade is too steep or abrupt, air will pass over the vent instead of through it. Since the vent will expose a portion of the southern facade, additional solar heat gain will result. Dense vegetation is effective in reducing solar heat gain as well as directing the flow of air into the vent.

GRADUAL SLOPE
TO VENT . . . . .

The number of vents required for a structure is determined by the square footage of the home, wall placement and building form. The placement of interior walls is extremely important in achieving efficient air flow. The better the air flow is, the fewer vents that are required. The following three examples of placement show both good and bad air wall movement patterns.

When a wall is located here it creates dead air spaces allowing stagnant air to build up.

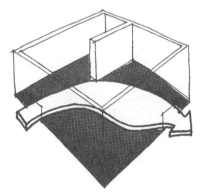

This wall configuration will work if additional vents are positioned as shown. The additional vents solve the problem by allowing dead air to escape.

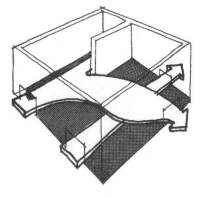

The wall in this location, allows good air circulation with no dead air spaces.

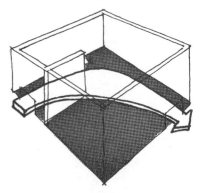

MODERATE/STEEP SITES
NORTHERN EXPOSURE

2 Moderate/Steep Sites, North Exposures

Ventilating naturally is difficult when a slope blocks prevailing summer breezes. Therefore HVAC systems, hoods, and exhaust fans may be needed to circulate and usher stagnant air outdoors.

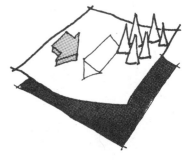

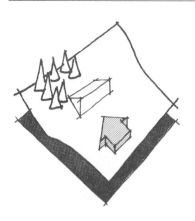

3 All Sites, Southern Exposures

Southern facades allow natural ventilation to work at capacity. To optimize natural ventilation, air should flow from the windward to leeward side. The size and placement of both intake and exhaust vents is critical. Due to the natural rise of hot air and fall of cool air, this requires the placement of intake vents low on the windward side and exhaust vents high on the leeward side.

OBSTRUCTION

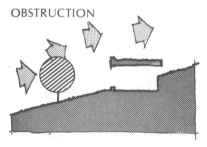

GUIDANCE

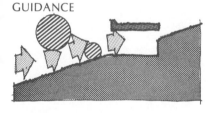

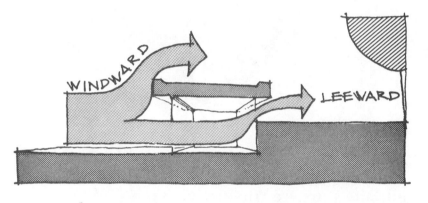

DEFLECTION

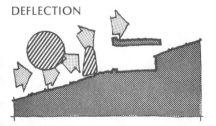

FILTRATION

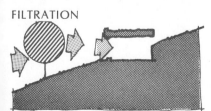

Plants may be used, in conjunction with landforms and architectural materials, to positively alter the airflow over the landscape or through structures for ventilation. Basically, plants control wind through one or any combination of the following methods: obstruction, guidance, deflection, and filtration. The key to successful wind control, through the method of vegetative planting, is the correct placement of plants. Although plants control wind effectively it must be remembered that plants as natural elements are not always predictable in their size, shape, and growth rate, and consequently, in their absolute effectiveness.

The following graphic illustration of venting utilizes all of the techniques mentioned previously, with the prevailing breezes perpendicular to the structure.

Dense hedges will deflect air into the structure.

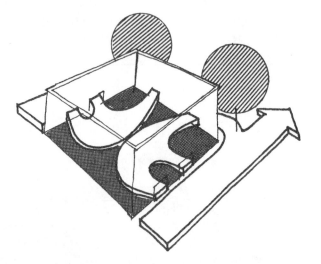

The trees filter wind and the hedges deflect air into the structure. The negative pressure formed by the hedge on the other side will draw the air out.

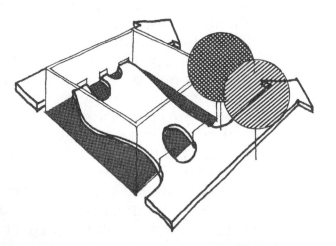

Trees and hedges which deflect wind create a
negative pressure which draws the air through the
structure.

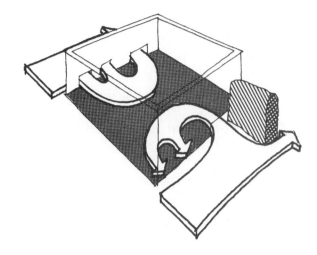

Hedges deflect and trees guide wind into the
structure.

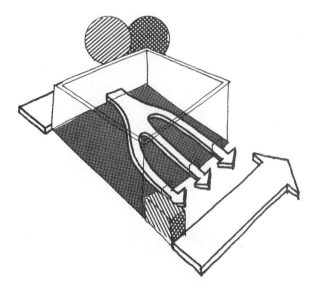

4 Moderate, Steep Sites/ East, West Exposures

The best methods of capturing wind for an east or west facade are landforms and cowls. A cowl is a device on an exterior vent pipe which acts as a weather vane and directs the exhaust opening downwind. Creating an air pocket with terraces and vegetation will channel breezes into the structure. Vegetation must be dense enough to deflect, not filter, winds into the structure. A mixture of dense hedges and trees will trap wind at all necessary levels.

A stationary or revolving finned cowl, based on natural ventilation principles, will effectively remove stagnant air from a structure. The cowl is installed to deflect the wind away from the vent opening. The revolving cowl is designed to turn or rotate with direction of the wind insuring negative suction on the leeward side. Warm air is removed through the exhaust pipe, and fresh air is introduced near the floor.

WINTER INFILTRATION

All four sites, slightly sloping, moderately sloping and steeply sloping, and orientations have been analyzed with respect to winter winds. The matrix below will help organize and identify specific conditions.

	NORTH	SOUTH	EAST	WEST
FLAT				
MODERATELY SLOPING	1	2		1
STEEPLY SLOPING, TOP				
STEEPLY SLOPING, BOTTOM				

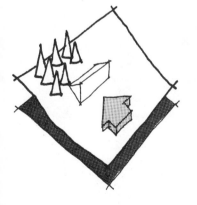

1 North/West Facades

North and west exposed facades are subject to harsh northwest winter winds which are undesirable and must be buffered.

Windbreaks can lower infiltration rates by effectively lowering wind velocity around the home. It is important to use vegetation which will yield a dense growth and grow to a height equal to or greater than the shelter. The maximum distance from shelter to windbreak should not be more than 5 times the home height.

Caution should be exercised in selecting the species and location of the planting when considering vegetation as a windbreak. Coniferous trees function better as windbreaks than deciduous trees because of their dense growth year round. Their shallow root system will not interfere with sewer lines and the drain tile of earth shelters. Root systems of deciduous trees extend as far as their outermost branches and therefore should not be located where they could interfere with the previously mentioned subsystems.

2 South/East Facades

Although out of reach of direct northwest winds, southern and eastern facades placed at the bottom of a slope will be chilled by cold air falling down the slope. Therefore, fences, walls, or any thick branched plants should not be installed at the bottom of the slope, as they would trap the cold air between themselves and the structure. The placement of a windbreak above the structure may alleviate and buffer the cold air currents falling downward.

The nest does not remain so small and simple a structure. It is soon extended into a multistoried building. We humans start our houses at the bottom—these builders add story after story from the top downward, suspending each new comb from the one above by columnar supports. They also extend the combs laterally. As the interior structure expands, the outer walls, too, must be enlarged. Parts that are too close to the center are pulled down, and new ones, providing new space, are added. . . . The stem from which the nest is suspended and the columnar supports are made from the same material as the thin, almost flimsy outer cover. Considerable load-bearing strength is achieved by aligning all wood fibers longitudinally —just as tendons of muscles derive their immense toughness from the fact that all the fibers of connective tissues are aligned parallel to each other in the direction of stress.

Karl Frisch, Animal Architecture (1974) p. 61

The Wasps Nest

3 STRUCTURAL SYSTEMS

Corrugated Steel Culvert

All-Weather Wood Foundation

Heavy Timber

Reinforced Concrete

The structural system of an earth shelter must resist earth pressures, the pull of gravity, and other manmade and natural forces including shock loads, winds, and earthquakes. In addition to resisting these forces, the structural system gives form to the building and creates architectural space within. When selecting a structural system, one must be aware of the building form and interior space. Developing an awareness of the potentials and limitations of a structural system will ultimately lead to a more responsive and successful design.

This chapter deals with structural and nonstructural elements of the earth-sheltered home. It examines potentials and limitations offering solutions to specific problems. Advantages and disadvantages of structural systems and materials are discussed with different viewpoints presented.

Corrugated Steel Culvert

Corrugated steel culverts, traditionally used only for drainage ditches, spanning creeks, and roadwork, are being used successfully in earth sheltered architecture.

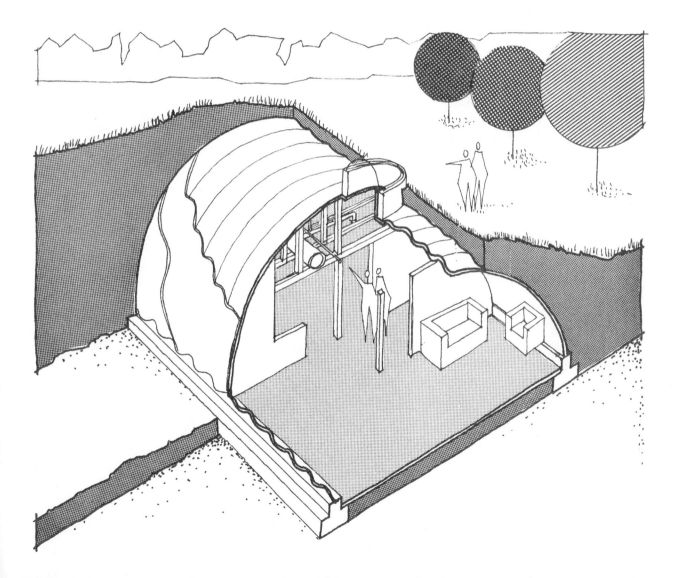

The steel arch of the culvert utilizes the "eggshell" principle, that is, ultimate stability of the structure depends on it being continuously loaded with earth. The culvert structural system is easy to erect, lightweight, economical when considering cost and materials, and efficient in loading.

EGGSHELL PRINCIPLE

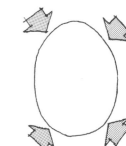

ERECTION PROCEDURE

To erect a culvert a level area on a site is first required.

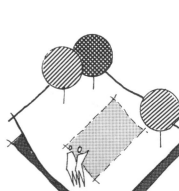

The foundation, which basically consists of two concrete trenches set parallel to each other, is poured next. The steel culverts are then bolted together in eight foot sections and the bases of the culverts are set into the trenches. This allows for any shifting within the earth and lets the culvert shift along with it.

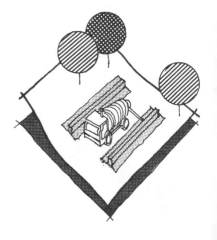

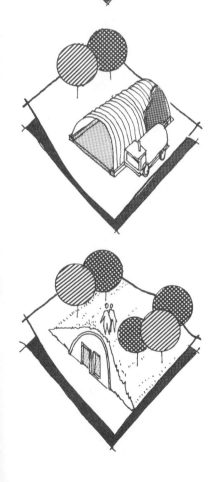

The floors are poured next and the insulation is applied to the walls usually in the form of polyurethane foam. A coating of fibrous asphalt is liberally applied to the outside walls once the culvert is totally assembled. After the asphalt has dried, multiple layers of mylar plastic film are put over the coating and then backfilled with sand to prevent puncturing. If overhead natural light is desired, holes are cut in the roof using a welder's torch and skylights are installed. The culvert is then ready for backfilling.

Backfilling can be completed after drain tiles are laid to carry subsurface water away from the home.

PROBLEMS AND ANSWERS

As with almost any innovative building idea some disadvantages will exist. No real building system exists since using culverts for homes is new. The culvert home is not built with a standard "kit of parts" which is available for above ground homes. This "kit" should include such items as doors, windows, skylights, and a choice of interior wall finishes. Consequently, building costs for construction may be higher due to unfamiliarity with the structure by contractors and more complicated design and construction techniques. However, costs should decrease as the building system is developed and more component parts are made available.

Culverts have severe interior space limitations because of their curved form. Entries and exits are restricted to the ends of the culvert, affecting home orientation with regards to site slope. The following are specific construction problems with steel culvert earth shelters and the solutions.

Problem

A problem arises when connecting a standard 2x4 stud wall or rafter ceiling to a curved metal surface such as a steel arch.

METAL BRACKET, TAC WELDED TO CULVERT
STEEL CULVERT
2x4 STUD BOLTED TO ARCH
DRY WALL FINISH

STEEL CULVERT
HEADER BOLTED
RAFTER
METAL BRACKET

DRYWALL

Problem

Since styrofoam or other rigid insulation may not conform to the curved steel arch, polyurethane or other foamed on insulation must be used. Polyurethane which emits toxic fumes when burning, is required by some state building codes to be enclosed or covered.

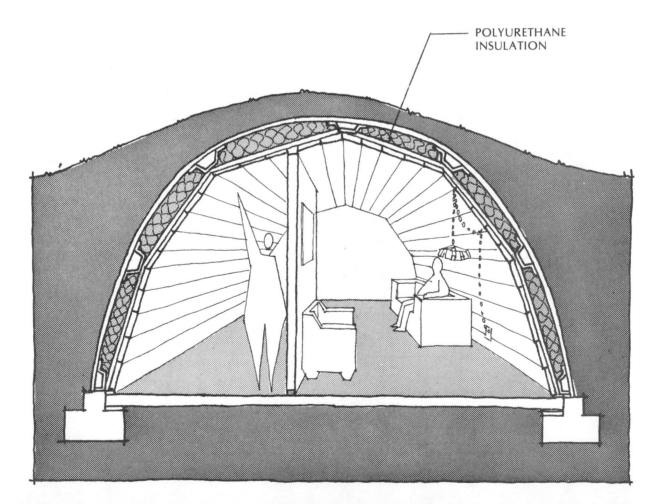

POLYURETHANE
INSULATION

ADVANTAGES

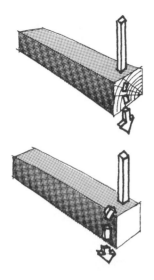

Steel beams, columns, and arches, because of the great strength of the material, are ideally suited for earth sheltered construction. Steel is durable and if kept dry will last a long time.

Leaking, a potential problem with wood or concrete is nonexistent in steel because of the material's high density and resultant watertightness.

DISADVANTAGES

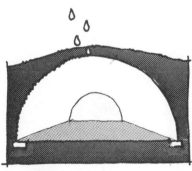

Steel rusts and will deteriorate quickly if not properly protected from water corrosion. Careful waterproofing and corrosion techniques must be taken to prevent damage to the material.

WATERPROOFING MEMBRANE

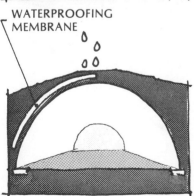

Steel must be covered with a fireproof material because it melts under high temperatures, typical of building fires.

All-Weather Wood Foundation

The All-Weather Wood Foundation System
(A-WWFS) is not only finding acceptance in above
ground building, but more importantly, finding
new potentials and applications in earth sheltered
construction.

Buried treated wood has been used successfully for
many years in railroad ties, utility poles, bridge
timbers, and high rise building piles.

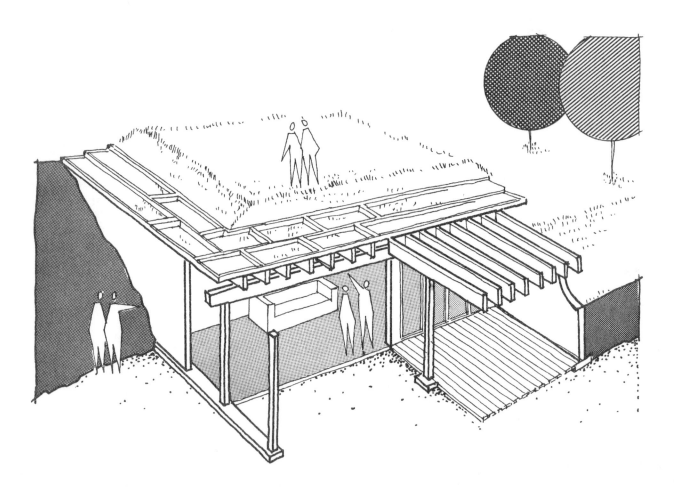

ADVANTAGES

A-WWFS have many advantages. Construction scheduling can be simplified since no mason is required for erection. A-WWFS can be erected in extreme temperatures because no mortar is used. Shop fabrication of foundation wall sections allows assembly on the site which reduces construction labor costs. Wood foundations will not crack because they are durable and strong. The plywood and stud construction can flex under earth pressures thereby maintaining structural integrity. Since cracking, a common problem in concrete block construction, is eliminated wood basements remain dry. A wood wall is easier to insulate and has a lower rate of thermal conductivity than a concrete wall.

It is considerably easier to apply interior finish to a wood wall as opposed to one of concrete block. Furring strips do not have to be attached to create a nailing surface for paneling or wallboard. Plumbing and wiring is as simple to install as in conventional above ground stud construction. Partitions and built-ins are also quickly installed.

A-WWFS have been approved by FHA/HUD and by the Farmers Home Insurance Administration for federal insurance mortgage programs.

DISADVANTAGES

Wood foundations can not withstand extreme earth pressures such as those existing at the base of a large slope. Reinforced concrete or concrete block is best suited for such a site. However, if the site is at the top of a steep slope, where the earth pressures are less, then a wood foundation is suitable.

The depth of an earth cover on a wood structure should remain shallow because of the limited load carrying capacity of ceiling rafters and stud walls. The load carrying capacity is the maximum amount of weight that can be carried by a structural member such as a beam or column.

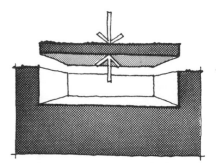

1

2

3

4

Shear is the tendency of a material to rip, tear, or separate under a force or load. Extensive nailing and gluing is required to develop resistance to high shear forces.

Wood foundations require a well-drained site to remain dry and to avoid uplift from subsurface water and ground pressure. This uplift can be demonstrated by submerging a glass beaker in a pail of water and then releasing the beaker. The rise of the beaker is similar to that experienced by a wood foundation under excess water pressure. Chemical treatment of the wood is required to avoid decomposition.

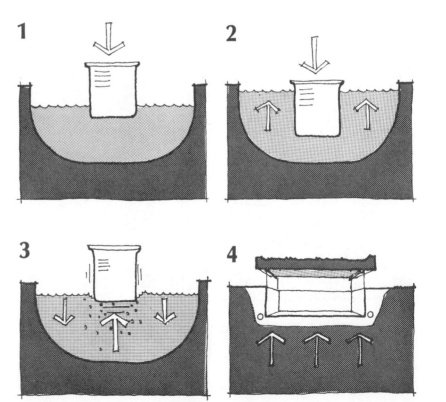

1

2

3

4

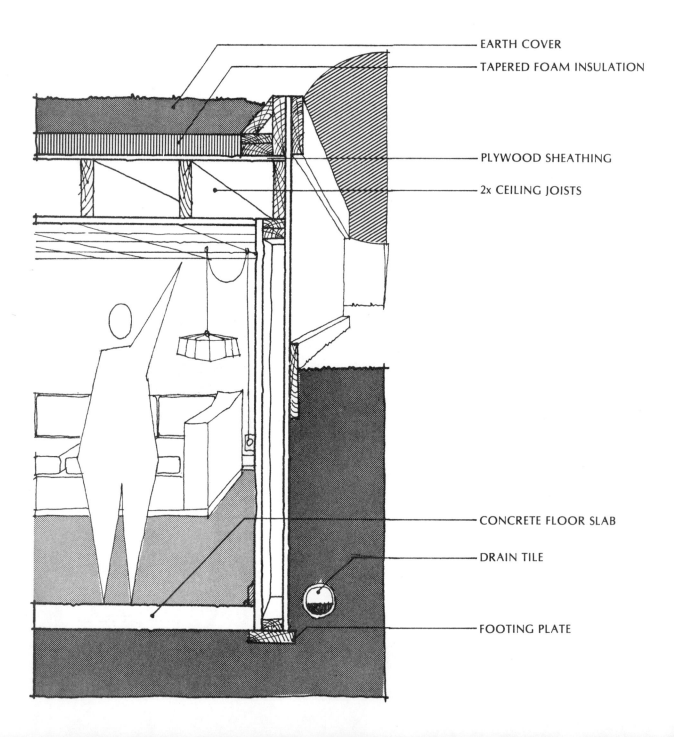

EARTH COVER

TAPERED FOAM INSULATION

PLYWOOD SHEATHING

2x CEILING JOISTS

CONCRETE FLOOR SLAB

DRAIN TILE

FOOTING PLATE

2x EXTRA STUDS. STUD AND TOP PLATE SIZE AND SPACING VARIES WITH HEIGHT OF BLACKFILL, SOIL PRESSURE AND VERTICAL LOADS . . .

TREATED PLYWOOD LAP TO COVER HALF OF FIELD APPLIED TOP PLATE . . .

TREATED PLYWOOD APPLIED WITH FACE GRAIN PARALLEL OR PERPENDICULAR TO STUDS. THICKNESS DEPENDS ON GRAIN ORIENTATION, HEIGHT OF FILL AND SOIL PRESSURE . . .

TREATED 2x BOTTOM PLATE END NAILED TO STUDS

FOOTING PLATE

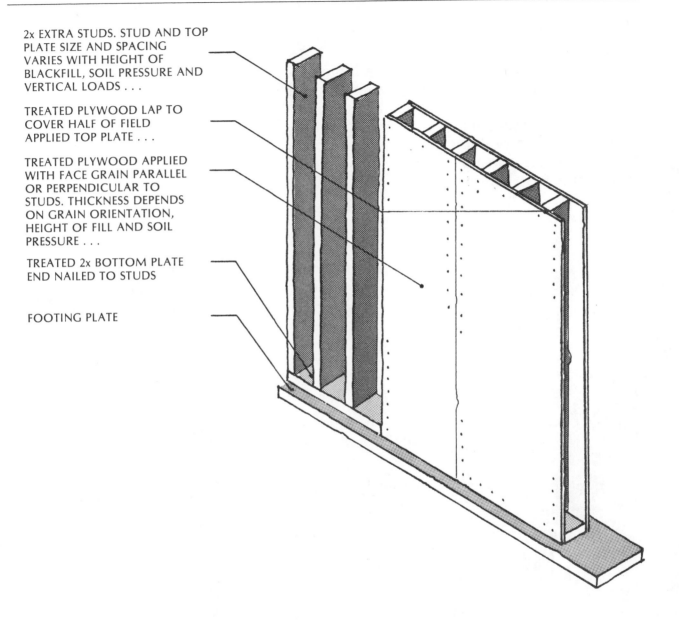

Heavy Timber

Heavy timber structural systems use large wood members as their primary support elements. Heavy timber has been successfully used in schools, churches, commercial and industrial buildings, residences, farm buildings, highway and railroad bridges, towers, theater screens, ships, and military and marine installations.

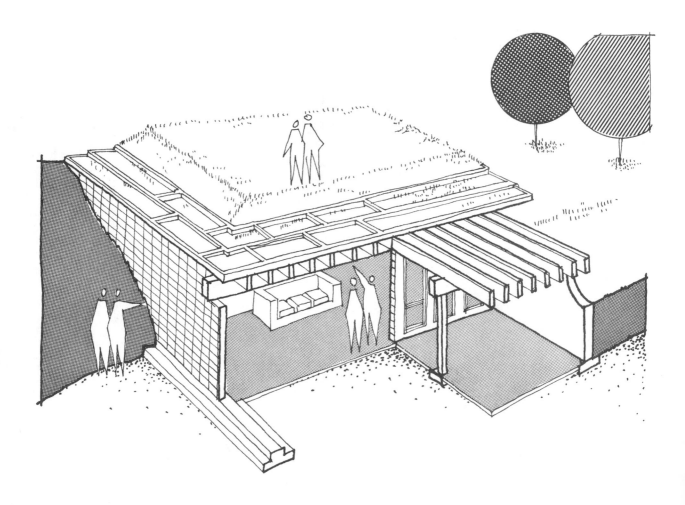

In the past timber was the predominant structural material in almost all buildings. However, due more to economics than performance capability, heavy timber is now being combined with such materials as steel, aluminum, masonry, and concrete. A heavy timber earth-sheltered home will usually combine wood, steel, and concrete.

ADVANTAGES

When properly exposed and treated, timber can be a visual amenity, adding beauty to the home. Therefore, it is desirable not to drywall, wallpaper, or paint wood surfaces.

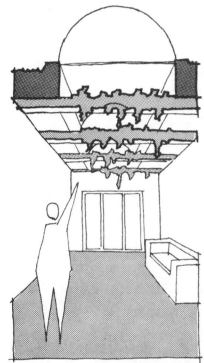

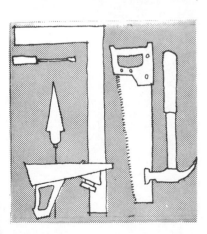

Timber is readily available in most areas and can be erected by standard carpentry crews. Expensive construction machinery and specialized workers can be eliminated in most cases.

Heavy timber, although combustible, has a high fire rating. During a fire initial charring by heat and flames creates a fire retardant coating on beams, columns, etc., rendering them more fire resistant.

DISADVANTAGES

Interior columns must be introduced into a home since wood cannot span great distances under a heavy load. Elongated building plans require minimal or no interior columns while square plans may require many.

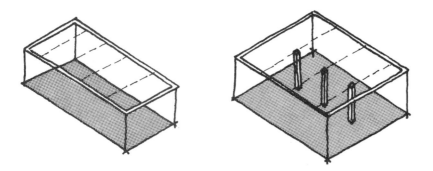

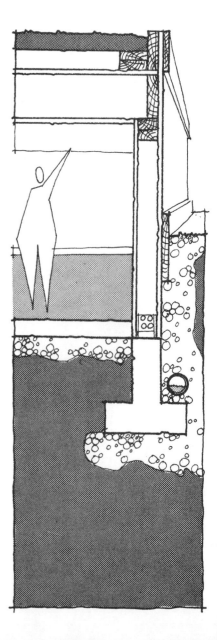

Since wood is a relatively lightweight structural building material, smaller footings and foundations are required than for concrete. Heavy timber structural systems required a well drained site to avoid uplift from subsurface water and ground pressure. This uplift principle is identical to the one discussed in the section on the All-Weather Wood Foundation system. Potential uplift due to underground water pressure makes it necessary to install more or larger drain tile, using gravel backfill for better drainage, or increase the size of the footings.

HEAVY TIMBER CONSTRUCTION DETAILS

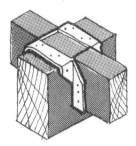

Rafter to Beam Connection

This rafter to beam connector is used for moderate and heavy loads. It provides a uniform fit where good appearance is desired. Rafters must be raised above the top of the beam to allow roofing deck to clear straps.

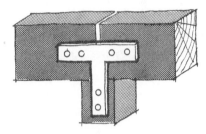

Beam to Wood Column

This steel T-plate connector is used to bolt beams to wood columns.

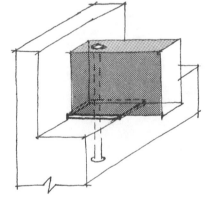

Beam to Wall Connection

This metal connector is used for beams less than 24 inches deep. A bolt embedded in the concrete wall secures the beam in place. Wood rot, a common problem resulting from moisture transfer from concret to wood, can be eliminated by placing a polyethylene moisture barrier between the two materials.

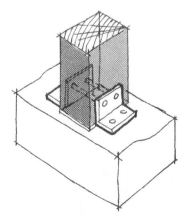

Column to Footing Connection

A column to footing connector is used to resist unusual loading conditions in addition to compression. Typical construction relies on pure compression to keep the column in place. This minimal condition is unsuitable for earth-sheltered homes.

CASE STUDY

The earth-sheltered home "Solaria" designed by
Malcolm Wells uses heavy timber in a unique and
interesting way.

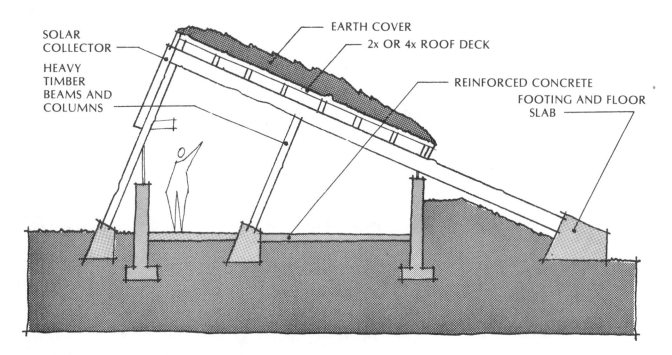

SOLAR COLLECTOR

HEAVY TIMBER BEAMS AND COLUMNS

EARTH COVER

2x OR 4x ROOF DECK

REINFORCED CONCRETE FOOTING AND FLOOR SLAB

The configuration of this heavy timber structural
system has many advantages. The triangular form
created by the ground line, slanted columns, and
main roof beam is very efficient under loading.
The triangle is a very stable form. The sloped
southern facade makes mounting solar collectors
easy since no additional structure or framing is
required to tilt the collectors. The sloped roof of
the building creates an interesting interior space.
Ponding of water, a potential problem on a flat
roof, is eliminated.

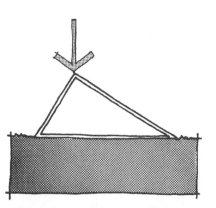

Reinforced Concrete

A concrete structural system is the most widely
used in earth-sheltered homes. It is characterized
by reinforced concrete footings, floor slabs,
foundations, reinforced concrete block bearing
walls, and prestressed precast concrete roof planks.
The primary materials are discussed.

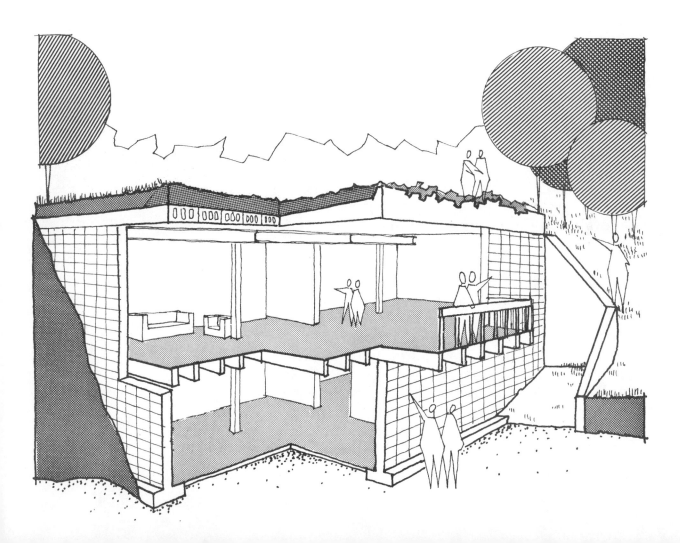

REINFORCED CONCRETE

Advantages

Reinforced concrete is durable, strong, and fire resistant. The durability of concrete is demonstrated by its ability to resist rot, rust, and decomposition by the elements. The combination of steel and concrete is extremely strong and ideally suited for two-story earth shelter design. Due to the non-flammable properties of concrete, it will not collapse during a fire. Due to its great density, concrete will act as an efficient integral passive heat storage system in the home.

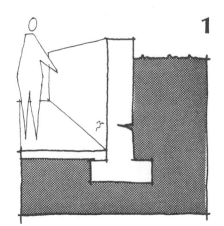

Disadvantages

The walls of reinforced concrete can crack. Reinforcing will prevent the wall from separating but will not prevent cracking. The cost of using concrete can be driven up considerably if the structural system requires large amounts of steel reinforcing. Although cracking may not be serious, it can lead to other problems such as leaking.

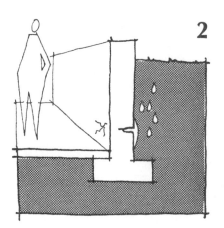

Although concrete is dense it is not impervious to moisture. This characteristic plus any cracking which may develop can lead to serious leaking. A variety of moisture barriers and waterproofing techniques have been developed which render water problems insignificant. A waterproofing material with the ability to span cracks should always be selected. Concrete holds residual moisture for several months after construction which cause temporary humidity problems in the home.

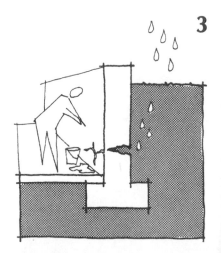

Concrete is extremely heavy and requires larger footings and foundations than heavy timber, steel arch, or All-Weather Wood Foundation structural systems.

Concrete, if not properly insulated, will act as a heat drain and actually draw heat away from the home during winter months. This is not a serious problem because a variety of very effective insulations are available.

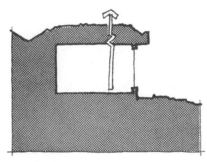

HEAT LOSS

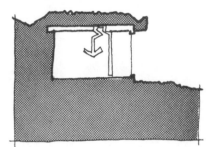

INSULATION PREVENTS HEAT LOSS

REINFORCED CONCRETE BLOCKS

Advantages

Premanufacturing concrete block in mass quantities has produced a wide assortment of shapes, patterns, and textures. Concrete block, in terms of aesthetics can provide a variety of forms since it can be assembled into large, complex shapes. The material is also readily available in most areas, which lowers cost and reduces delivery time.

Concrete block offers poor resistance to heat transfer and must be properly insulated. Insulation can be poured into the hollow cores of concrete block, thus reducing the amount of insulation required on the interior and exterior surfaces.

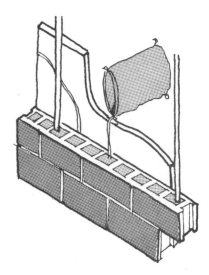

Concrete block is also durable, moderately strong and fire resistant. A concrete block wall contains mortar joints which increase the potential for cracking and leaking.

Disadvantages

Concrete block is heavy, requiring large footings and foundations.

Concrete block walls are poor moisture barriers, and will leak if not carefully waterproofed.

Special reinforcing is required to prevent block walls from cracking under large earth pressures. Pilasters, steel reinforcing in grout filled block cores, and kicker walls are three common methods to limit cracking.

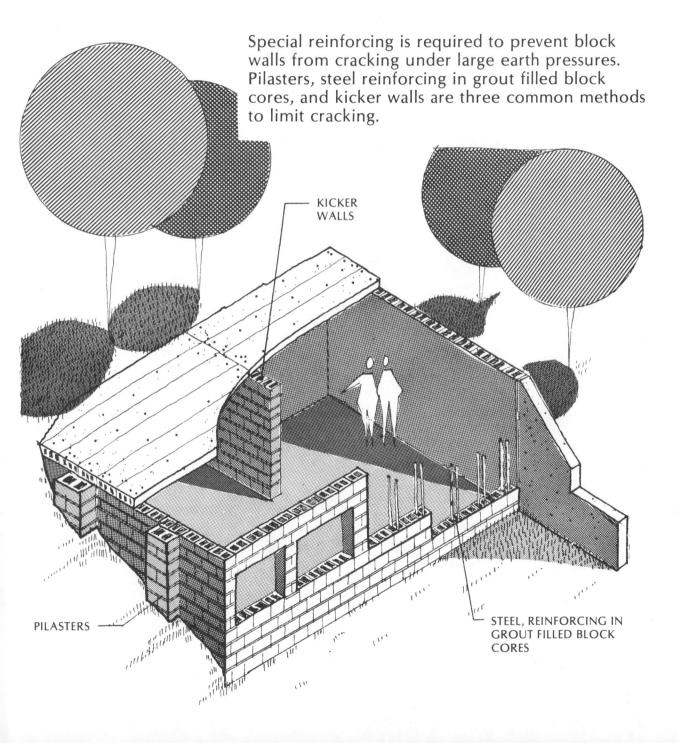

KICKER WALLS

PILASTERS

STEEL, REINFORCING IN GROUT FILLED BLOCK CORES

PRECAST PRESTRESSED CONCRETE

Advantages

Precast concrete planks, normally produced in two to eight foot widths, have found wide acceptance in earth-sheltered housing. They can be assembled to form wall sections, but more commonly, are used as structural roof elements. Concrete, although strong in compression, will easily fail in tension. Steel, however, is able to resist high tensile forces. The combination of concrete and steel produces a precast prestressed concrete plank which creates an ideal roof material.

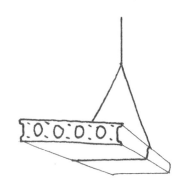

- COMPRESSION   - TENSION

- PRECAST CONCRETE PLANK
  COMBINES TENSION
  & COMPRESSION
  MEMBERS . . .

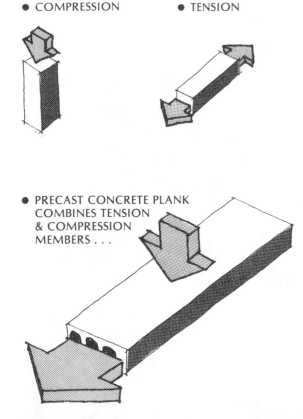

The voids in concrete planks can be easily converted into plumbing, electrical, and venting raceways.

Prestressing introduces compressive strength into the material which compacts the cement binder and aggregate to the extent they become impermeable to the penetration of water.

Disadvantages

The economy of concrete planks is dependent upon the distance between the building site and the plank factory. Long distances will increase transportation costs and potentially delay construction.

Planks are usually limited to flat linear shapes, thereby restricting building form.

If joints between planks are not properly sealed and waterproofed leaking could develop.

PROBLEMS AND ANSWERS

Problem

The appearance of unfinished concrete ceiling planks may not be desired within the home.

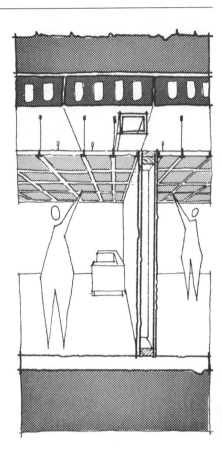

If the appearance of concrete or precast ceiling panels is not desired, a suspended acoustical ceiling can be installed. The void created by the false ceiling can be used to hide power conduit, water lines, and HVAC ducts.

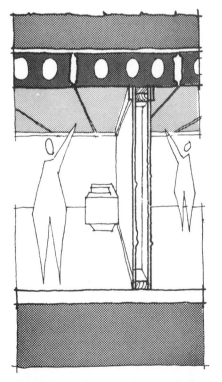

Precast concrete planks are factory produced which insures a high quality finished product. The production process yields planks of uniform size and shape, with smooth finishes. A plaster coating can be easily sprayed onto the planks for a textured finish.

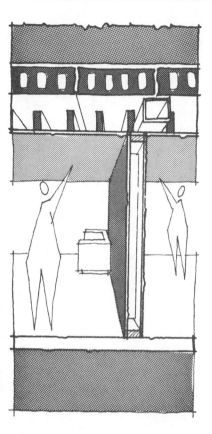

A conventional drywall ceiling matching interior walls is possible by adding ceiling rafters.

CASE STUDY

The two-level earth-sheltered home is compact. It encloses the greatest amount of floor space with the least amount of exterior surface area. This saves energy due to a reduction in heat loss.

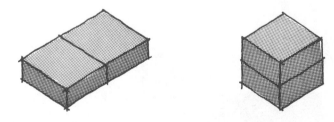

If greater soil depths are desired on the roof, then a center support for the roof planks should be provided. A wide flange steel I-beam bearing on steel columns provides the necessary support.

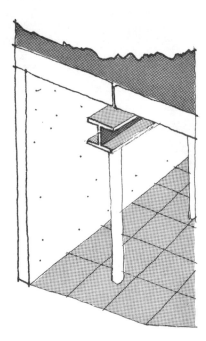

The second floor system, usually of 2x10 or 2x12 floor joists carry and conceal utilities such as heating ducts, water lines, and electrical cables to both floor levels. Since the floors are stacked, half as many concrete roof panels are needed as in a single level home.

Two-level earth-sheltered homes require larger foundations and thicker bearing walls on the first floor because of increased lateral soil pressures.

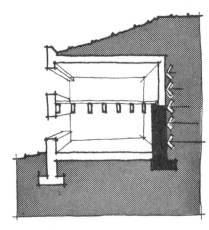

A two-level earth-sheltered home can still offer a view if built on a flat site.

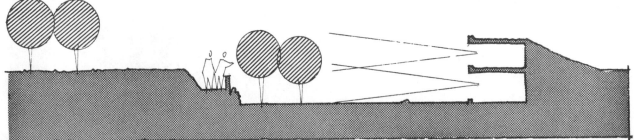

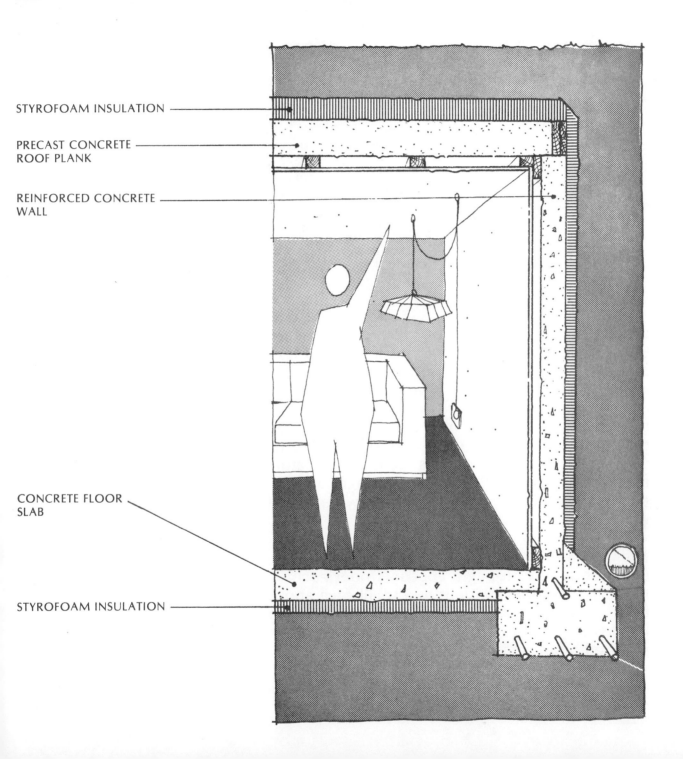

STYROFOAM INSULATION

PRECAST CONCRETE
ROOF PLANK

REINFORCED CONCRETE
WALL

CONCRETE FLOOR
SLAB

STYROFOAM INSULATION

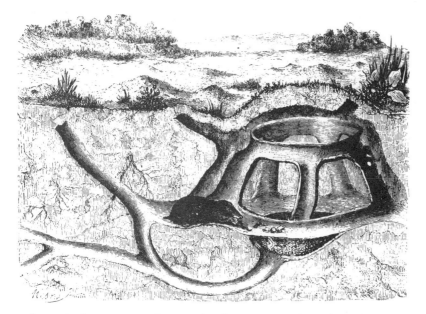

When it burrows through the ground, it loosens the soil with its snout and two front legs and then partly presses the loosened soil against the side of the tunnel with a turning movement of its body like that of a drill, and partly pushes the dirt toward the back with its hind legs. From time to time it pushes through to the surface and throws up a molehill. These molehills serve as ventilators for the network of subterranean passages and enable the mole to reach the surface whenever it wants to . . .

The mole's living sphere is, however, not so uncomfortable as it might appear. Under each molehill, the tunnel widens to a small cave about twenty centimeters in diameter; this is comfortably lined with grass and dry leaves. These chambers are used by the animals for resting and sleeping . . .

The Mole

Karl Frisch, Animal Architecture (1974)
p. 248

4 BUILDING COMPONENTS

Weather Locked Entries

Soil Retainers

Insulation

Retaining Walls

Skylights

Windows

Waterproofing

Standard building elements or components have
been easily adapted to earth-sheltered homes. Each
component has an assortment of alternatives which
are interchangeable between the four structural
systems. These components are essential to the
success of an earth-sheltered home.

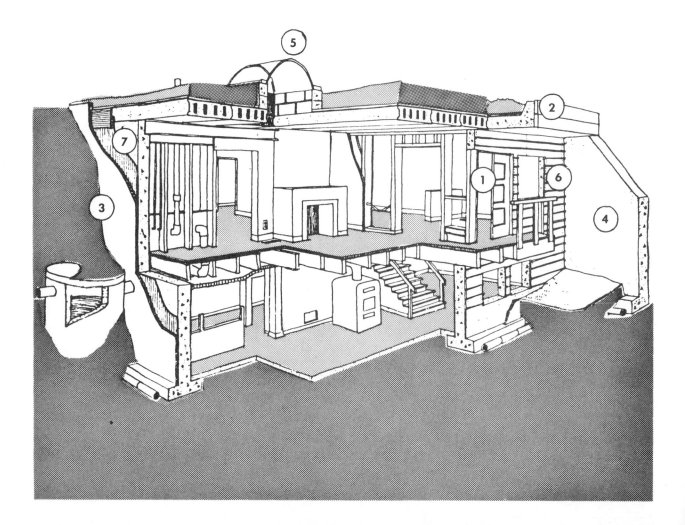

1 Weather Locked Entries

In many of todays residences the heat loss associated with entering and leaving a home is still a problem. Most people are familiar with the usual blasts of cold winter air that usher in family members and visitors. The opening of a non-weather locked door during subzero weather, just for a few seconds, can allow several hundred cubic feet of cold air to enter.

A simple solution to the problem has been the introduction of the air or weather lock. An entry with two doors, in which the first closes before the second is opened, minimizes the entry of cold air into the home.

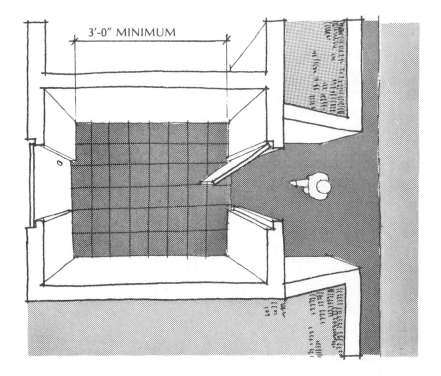

A specific type of weather lock is shown with each structural system. However, the concepts are interchangeable and can be combined in one home. In a weather locked entrance the outside door should have two inches of polyurethane insulation sandwiched between two sheet metal or wood layers. The doorframe should be sealed with magnetic weatherstripping. The inside door should be at least three feet from the outside door, be constructed of wood, and also be weatherstripped. Both doors should open inward.

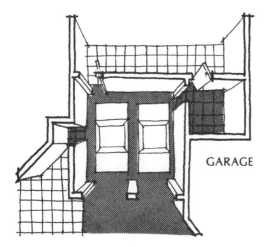

GARAGE

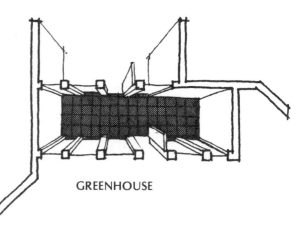

GREENHOUSE

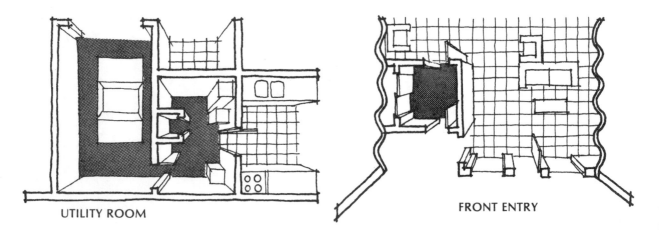

UTILITY ROOM

FRONT ENTRY

2 Soil Retainers

Once soil is placed on the roof of an earth-sheltered home, it should remain in place. Several devices, techniques, and special roof edge details have been developed for this purpose.

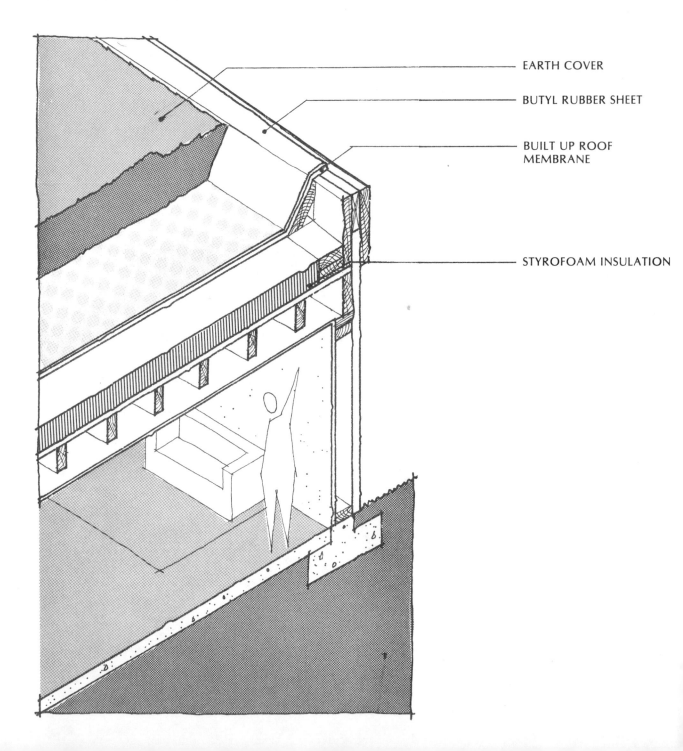

EARTH COVER

BUTYL RUBBER SHEET

BUILT UP ROOF
MEMBRANE

STYROFOAM INSULATION

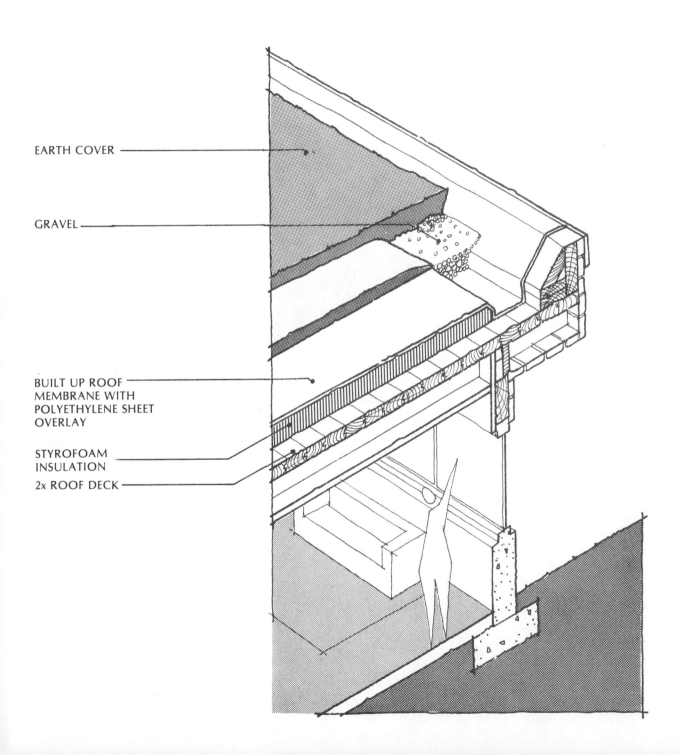

EARTH COVER

GRAVEL

BUILT UP ROOF
MEMBRANE WITH
POLYETHYLENE SHEET
OVERLAY

STYROFOAM
INSULATION

2x ROOF DECK

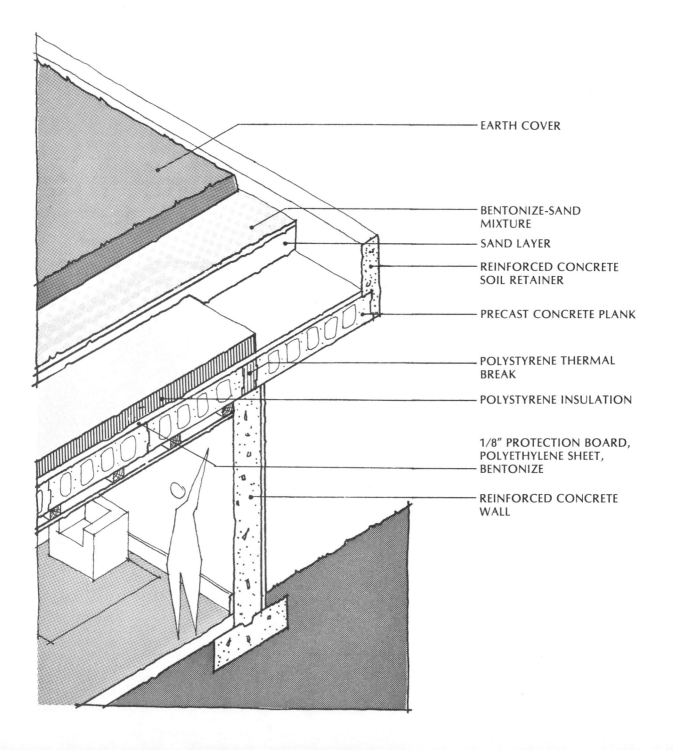

EARTH COVER

BENTONIZE-SAND MIXTURE

SAND LAYER

REINFORCED CONCRETE SOIL RETAINER

PRECAST CONCRETE PLANK

POLYSTYRENE THERMAL BREAK

POLYSTYRENE INSULATION

1/8" PROTECTION BOARD, POLYETHYLENE SHEET, BENTONIZE

REINFORCED CONCRETE WALL

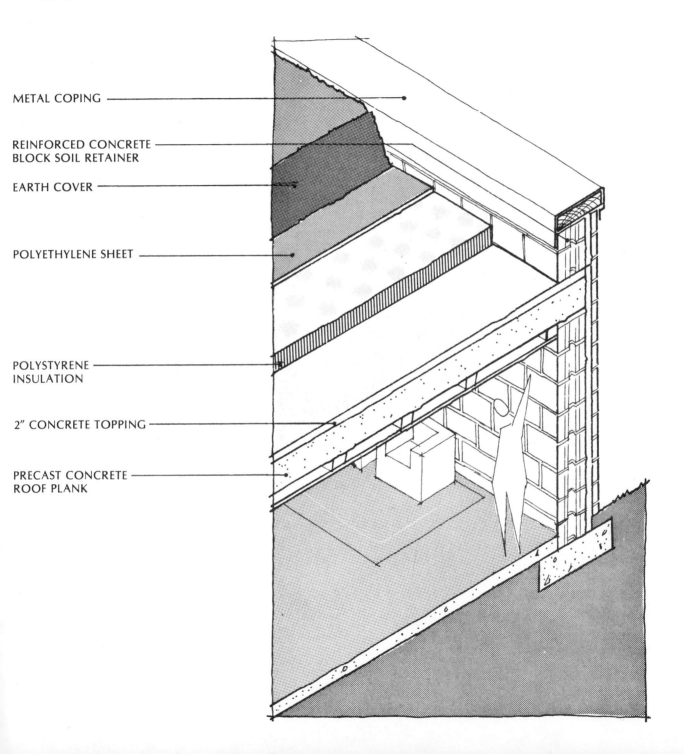

METAL COPING

REINFORCED CONCRETE
BLOCK SOIL RETAINER

EARTH COVER

POLYETHYLENE SHEET

POLYSTYRENE
INSULATION

2" CONCRETE TOPPING

PRECAST CONCRETE
ROOF PLANK

3 Insulation

Insulation is any material which reduces the transfer of heat from one area to another. All building materials have some insulating value, but the term "insulation" generally refers to products designed primarily for this purpose. Insulation helps keep heat in the building during cold weather and out during hot weather. Insulating materials are rated according to their "R" value or the materials resistance to heat transfer. A high "R" value indicates little heat transfer. Conversely, a low "R" value indicates high heat transfer. Local building codes and up-to-date government and private research studies of health hazards of various insulation materials should be checked prior to final selection. Following is a brief description of six basic types of insulation now available and their use in earth-sheltered homes.

Material	Resistence to heat transfer
Polystyrene—	R 3.4 to 4.5 per inch aged
Vermiculite—	R 2.2 per inch
Cellulose or wood fiber—	R 3.7 per inch
Fiberglass—	R 3.1 to 3.7 per inch
Polystyrene, pour fill—	R 3.57 per inch
Urea Formaldehyde—	R 5.0 to 5.5 per inch
Urethane—	R 6.25 per inch aged

POLYSTYRENE

Polystyrene, a rigid insulation, is suitable for a wide variety of applications. Most importantly, it is used for insulating concrete foundation walls. Extruded polystyrenes, styrofoam, have a high insulating value and are resistant to moisture absorption. Molded polystyrenes, beadboard have low "R" values and a low resistance to the passage of moisture. Polystyrene should be covered with a fire resistant material because of its flammable nature. Extruded polystyrenes have high insulation values when new because of the gases used in foaming. As it ages the gases escape and the "R" value decreases.

Styrofoam (a trademark name of the Dow Corporation) is the most highly recommended insulating material. It is generally blue or gray in color.

Other materials such as white beadboard or urethane soak up water and deteriorate. Insulating below the frost line with two inches of styrofoam is recommended for earth shelters. In the Midwest, East, and other rainy areas, it is probably best to insulate around the whole building. One inch styrofoam sheets can be used underneath the floor slab to reduce heat loss. Styrofoam is also used as a thermal break.

Polystyrene thermal breaks are commonly used to reduce direct heat transfer from an exterior concrete surface to an interior concrete surface.

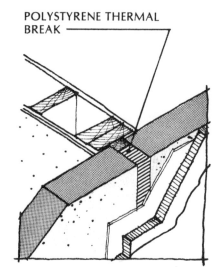

POLYSTYRENE THERMAL BREAK

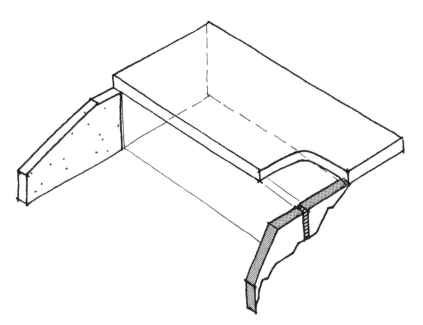

VERMICULITE

Vermiculite has a lower insulating value than most other insulations. Vermiculite is suitable for insulating ceilings when additional "R" value is desired. This insulation is extremely fire resistant and free flowing. It has been widely used for filling the cores of concret blocks to increase the "R" value of a block wall.

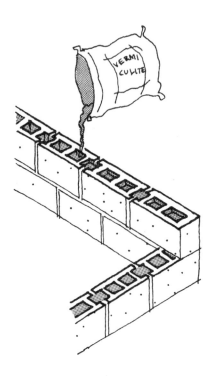

POLYSTYRENE POUR FILL

The use of polystyrene bead pourfill inside stud walls is a preferred alternative to fiberglass. It costs less than fiberglass, has a high "R" value, and when blown into a stud wall under pressure will not settle or shrink. Polystyrene and polyurethane insulation are government approved when covered with drywall and siding material.

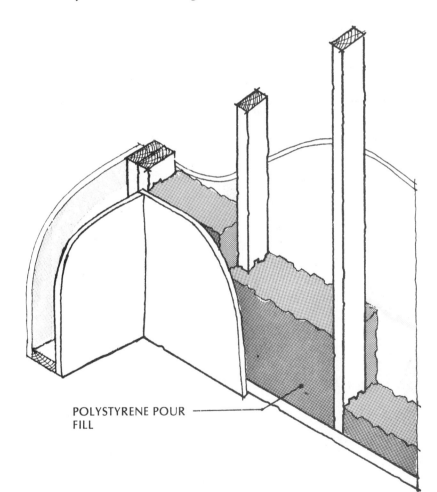

POLYSTYRENE POUR FILL

CELLULOSE OR WOOD FIBER

Cellulose is a fill-type insulation made from paper or pulp products that has been chemically treated for fire resistance. Commonly, it is blown into the stud walls of existing buildings and is also widely used for insulating ceilings.

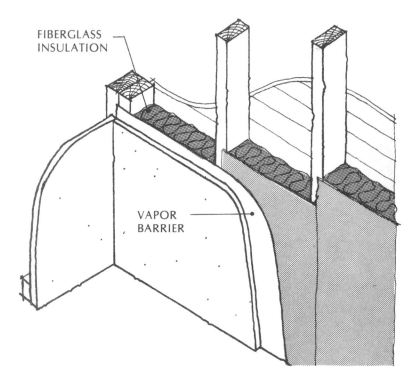

FIBERGLASS INSULATION

VAPOR BARRIER

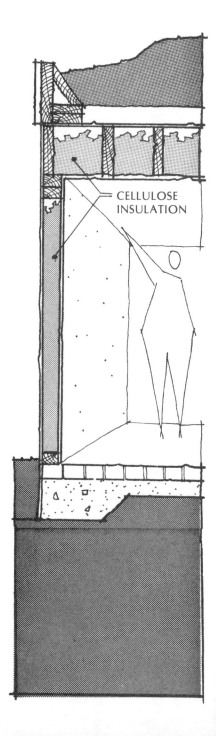

CELLULOSE INSULATION

FIBERGLASS

Fiberglass batt or blanket insulation is most commonly used in the walls of new construction. Blown-in fiberglass is used extensively for insulating ceilings. Blanket insulation is available with a vapor barrier backing, but a separate vapor barrier between the insulation and the inner wall is desirable in buildings where moisture is a problem.

UREA FORMALDEHYDE

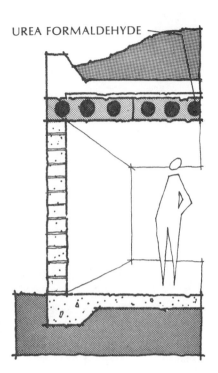

UREA FORMALDEHYDE

Urea formaldehyde is a relatively new insulation which is foamed-in-place and sold under a variety of trade names. It is ideal for insulating existing walls because the foam will flow into small openings and crevices. The material is fully expanded as it leaves the applicator and it solidifies within about one minute. Urea formaldehyde is highly resistant to moisture absorption and has properties which rodents normally avoid. It is also well suited for insulating the hollow cores of precast concrete panels.

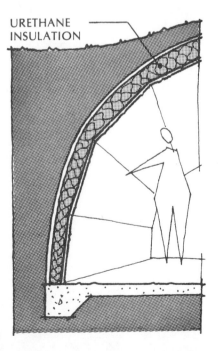

URETHANE INSULATION

URETHANE

Urethane has the highest insulating value of all the building insulations. Rigid sheets are available in a wide range of thicknesses and it may be sprayed onto walls or ceilings. Urethane should definitely be covered with a fire resistant material, as the fumes it emits during a fire are extremely toxic. Initially urethane has a high insulating value, but after a 10-15 year period this tends to decrease.

4 Retaining Walls

Earth-sheltered buildings usually require earth-moving in the form of excavating, backfilling, or berming. These actions require the use of free standing retaining walls to stabilize soil and reduce the erosion on steep slopes. A retaining wall, which holds back or retains earth between levels of earth, utilizes the weight of the wall plus the weight of the earth over the footing to resist overturning.

No one retaining wall is peculiar to any given site situation. It is recommended, however, that if a great deal of earth is to be supported as in a full story or greater retaining wall, then reinforced concrete or concrete block should be used.

Common retaining walls such as the gravity, cantilever, and rubble, in addition to other grade change methods, are illustrated below.

GRAVITY OR MASS RETAINING WALL

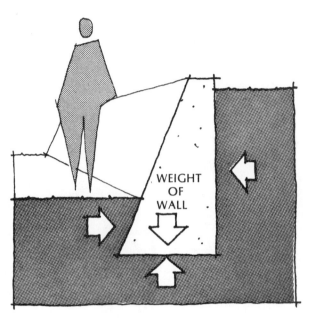

WEIGHT
OF
WALL

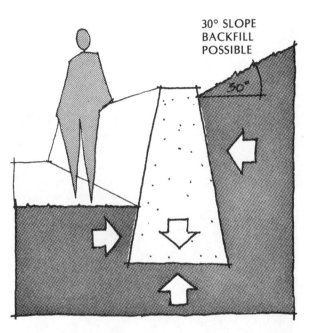

30° SLOPE
BACKFILL
POSSIBLE

30°

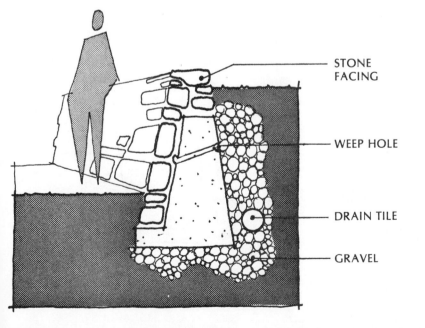

STONE
FACING

WEEP HOLE

DRAIN TILE

GRAVEL

A GRAVITY RETAINING
WALL DEPENDS ON ITS
MASS TO RESIST
OVERTURNING FORCE
OF BACKFILL

CANTILEVER "L" TYPE RETAINING WALL

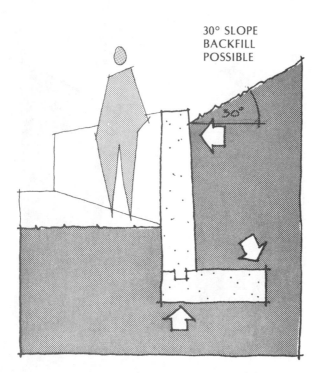

30° SLOPE
BACKFILL
POSSIBLE

30°

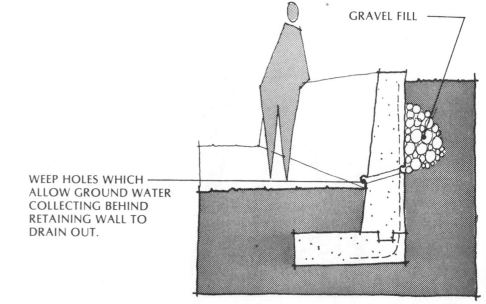

GRAVEL FILL

WEEP HOLES WHICH
ALLOW GROUND WATER
COLLECTING BEHIND
RETAINING WALL TO
DRAIN OUT.

CANTILEVER "T" TYPE RETAINING WALL

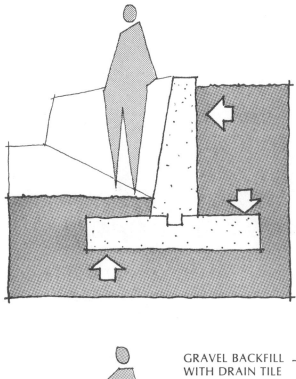

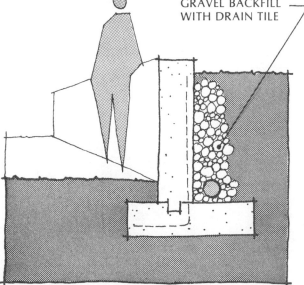

GRAVEL BACKFILL
WITH DRAIN TILE

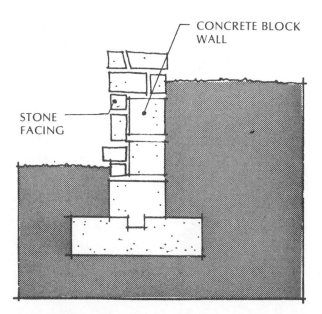

CONCRETE BLOCK
WALL

STONE
FACING

OTHER GRADE CHANGE TECHNIQUES

These methods are generally less expensive than the retaining walls previously discussed and offer the potential to be owner built.

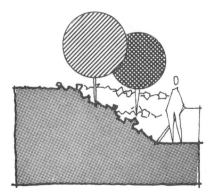

EARTH EMBANKMENT

STONE RIPRAP

REINFORCED EARTH

Over a period of many years these grade change techniques may not be as effective as concrete retaining walls due to the impermanence of the material and lesser anchorage.

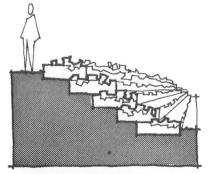

TIMBER OR CONCRETE
STEPPED CRIBBING

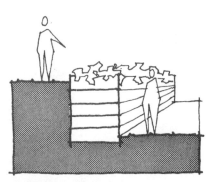

VERTICAL CRIBBING

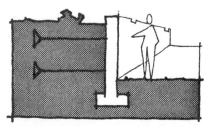

TIE BACKS

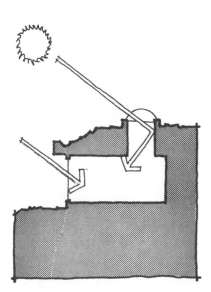

5 Skylights

In some earth-sheltered homes spaces may exist which are unable to be completely lit by conventional windows. Such rooms as bathrooms, corridors, and workshops, may require overhead skylights for daylighting. If skylights are properly selected, placed, sized, and constructed, these interior spaces can become very pleasant.

Since the use of skylights to illuminate interior rooms is essential, basic principles of daylighting are worthy of discussion.

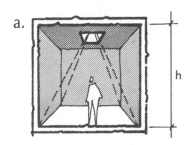

DAYLIGHTING PRINCIPLES

An important factor when sizing or selecting a skylight for a room is the room's floor to ceiling dimension. The taller the room is the larger the skylight should be if overall general illumination is desired.

a. 1/1 LIGHT INTENSITY
STRIKING FLOOR

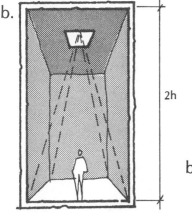

b. 1/4 LIGHT INTENSITY
STRIKING FLOOR

It also follows that the larger the skylight, the greater the light intensity for rooms of similar size.

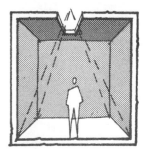

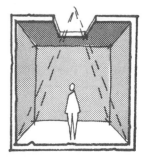

Deep well skylights, designed to control direct sunlight and allow only reflected light to enter a room, absorb much of the light before it reaches the floor plane. To prevent this, one must adjust the width and depth of the well or use a different daylighting device.

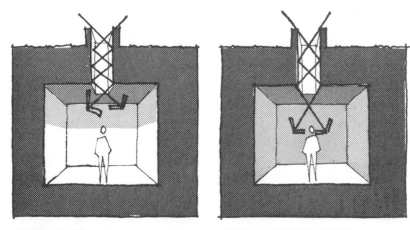

EXPECTATION   ACTUAL PERFORMANCE

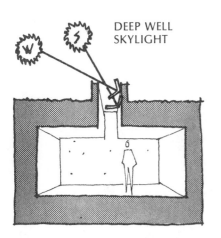

DEEP WELL
SKYLIGHT

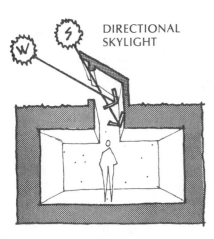

DIRECTIONAL
SKYLIGHT

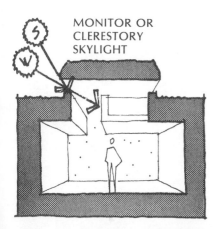

MONITOR OR
CLERESTORY
SKYLIGHT

Direct sunlight fades artwork, drapes, rugs, woodwork, and furniture. Although direct sunlight is essential for passive heat gain in the home during the winter it can cause problems such as overheating during the summer as well as fading. Therefore, when selecting a skylight the positive and negative factors must be considered. Several daylighting devices have been developed to manipulate and decrease the negative effects of direct sunlight.

The light penetrating into a room will be more diffused if wall surfaces within the skylight are textured. If greater reflectance is desired, then smooth or shiny surfaces should be used.

The glazing material in skylights is glass or plastic (acrylic). Plastic glazing and glass is usually clear, tinted, or transluscent. Glass also comes plain, tempered, or embedded with wire reinforcing. Plastic glazing is available in single or sandwiched plastic panels.

Skylights are poor insulating devices offering little resistance to heat loss. This poor insulating characteristic coupled with their usual locations in warm ceilings results in tremendous heat loss. To improve the insulating value of glass, it is recommended that a skylight be used with a minimum of one, preferably, if possible, with two dead air spaces.

OTHER DAYLIGHTING TECHNIQUES

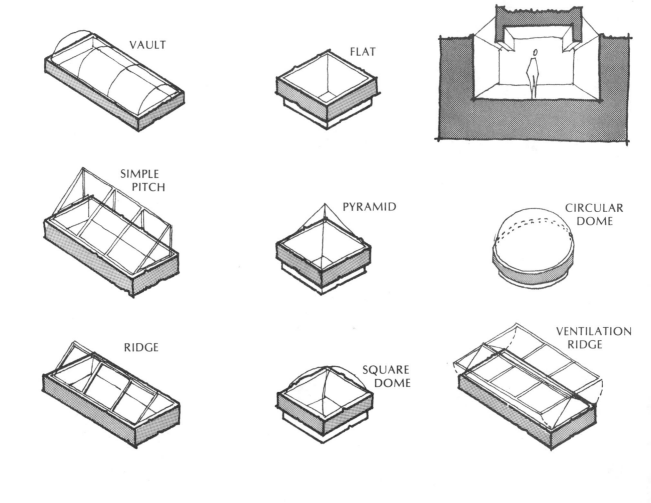

VAULT

FLAT

SIMPLE PITCH

PYRAMID

CIRCULAR DOME

RIDGE

SQUARE DOME

VENTILATION RIDGE

CORRUGATED STEEL CULVERT

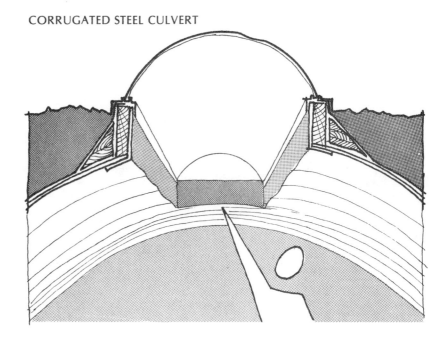

ALL-WEATHER
WOOD FOUNDATION

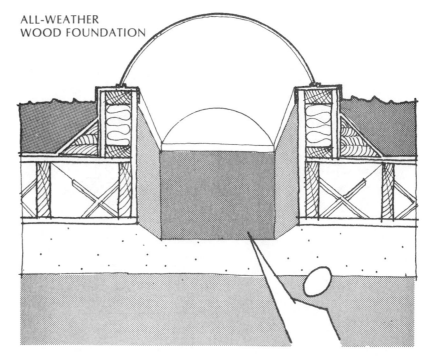

HEAVY TIMBER

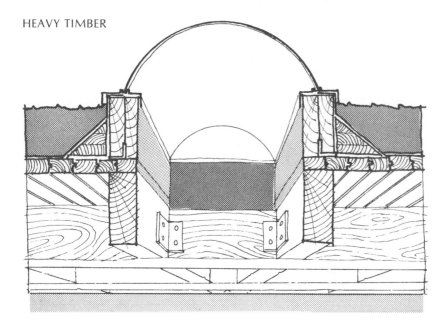

REINFORCED
CONCRETE

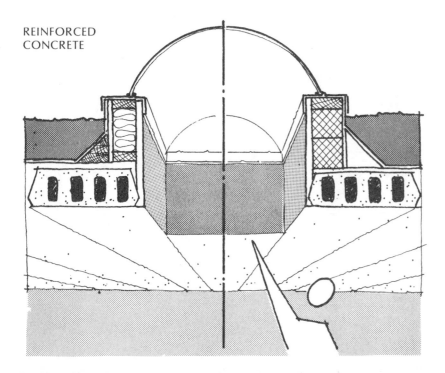

6 Windows

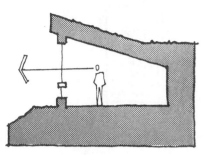

Earth-sheltered homes, as their above ground counterparts, require windows. In addition to letting in natural light, they provide views, allow for ventilation, and provide an exit during emergencies. Windows, which are essential to a pleasant interior space, offer protection from the elements while still allowing visual contact with the outdoors. Windows are required to be weather-tight when closed, have insulative value, and be free from condensation.

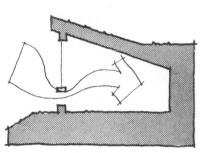

WINDOW LOCATION

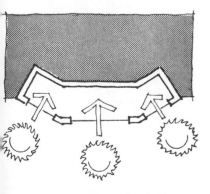

In keeping with the energy conscious nature of earth-sheltered homes, whenever possible, windows should be located on the southeast, south, or southwest exposures for maximum solar gain.

HEAT LOSS

Glass, because of its thin homogeneous nature, is a poor insulator and offers almost no resistance to heat flow. Recent technological advances have introduced double and triple glazed thermalpane glass which has a greatly increased insulating value. Thermalpane glass windows basically consist of two or three layers of glass separated by a 1/2 inch dead air space that functions similar to insulation by reducing heat transfer. The air spaces are vacuum sealed to eliminate condensation between the panes. For a minimum of heat loss triple insulating glass in conjunction with a storm window, separated by a four inch dead air space is recommended. Costs for such elaborate windows are high, but during the lifespan of the building, that extra added expense is normally returned in energy savings. Wood window frames should always be used as opposed to aluminum frames. Heat will easily pass through a metal window frame thereby reducing the overall insulating value of a thermalpane window.

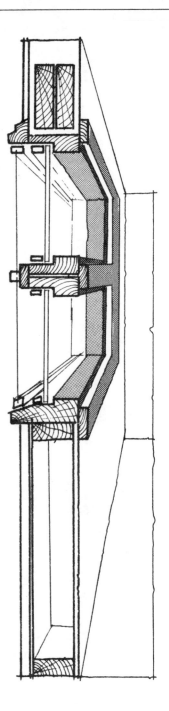

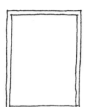

FIXED

0% VENTILATION

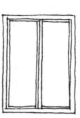

SLIDING

50-60%
VENTILATION

CASEMENT

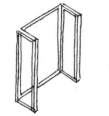

100% VENTILATION

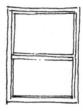

DOUBLE-
HUNG

50%
VENTILATION

AWNING

100% VENTILATION

HOPPER

100% VENTILATION

PIVOTING

100%
VENTILATION

All of the window types shown can be used in
earth-sheltered homes. However, a combination of
fixed triple glazed insulating and hopper windows
is recommended if available. Fixed windows allow
for maximum light and passive solar penetration.
They enclose large glass areas with a minimum
amount of seams, thereby reducing the potential
for air and moisture leakage. Hopper windows
placed low on the floor enhance natural
ventilation. They allow cool air to enter low,
forcing warm stagnant air near ceilings to vent out
through skylights or roof vents. (See summer
cooling section.)

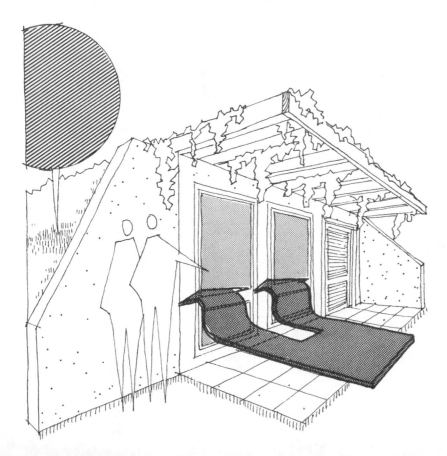

7 Waterproofing

Anyone who has ever lived in a home with a basement is all too familiar with the problems of high humidity and moisture condensation. Since most basements were never intended to be living spaces, thorough waterproofing measures were seldom made. In earth-sheltered homes, however, waterproofing is essential to prevent leakage and related problems such as water damage, staining, and dampness. An earth-sheltered home, if properly designed and waterproofed, will remain dry and problem free. Taking design precautions prior to waterproofing will hopefully avoid as many water problems as possible. This is preferable to coping with them after the home is completed.

Sites with high water tables, in flood plains, and in low areas should be avoided. These areas can be built in, but more extensive waterproofing and runoff control measures must be used with no guarantee of success.

Floor drains should be installed in courts or other sunken areas and cleaned regularly.

It is advantageous that ground slope away from an earth-sheltered home. This insures that excess surface runoff will not drain toward the house.

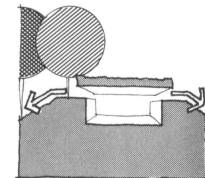

Prior to backfilling the walls, drain tile should be installed preferably at the roof and footing line to carry away subsurface water. Gravel, sand, or other free draining material should be used for backfilling.

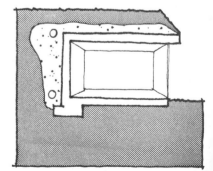

The success of any waterproofing system depends on the design measures (already) discussed, type of waterproofing material selected, and high quality control during installation.

Dampness due to capillary draw of ground moisture into block or concrete walls and leakage due to wall cracking are two problems that waterproofing techniques attempt to eliminate in earth-sheltered homes.

Several waterproofing techniques and the specific materials involved are illustrated and discussed below. This list is by no means all inclusive since this discussion does not attempt to explain every waterproofing product available. The criteria for a good waterproofing product is that it last, be easy to apply and inspect during application, be self sealing, and the ability to expand and bridge cracks.

Bituthene

Bituthene is a polyethylene coated rubberized asphalt. Bridging characteristics are reasonable. A roof waterproofed with Bituthene should have a slight slope for drainage since the material is not intended to be used under constant water pressure. This membrane should also be covered because ultraviolet rays induce deterioration of the polyethylene.

Butyl Rubber

Butyl rubber is a quality waterproofing material with good bridging characteristics. The integrity of the membrane depends on the correct sealing of the seams between sheets and that care be taken to avoid puncturing during backfilling. Water entering a hole in the membrane can travel laterally beneath the membrane and appear as a leak in a remote place. This makes locating the hole difficult and expensive.

Built-up Membranes

Membranes consist of layers of asphalt or pitch alternated with flat or fabric reinforcing. Four plies are considered minimum for waterproofing purposes for an earth-sheltered home. The fabric, usually a glass fiber, gives some mechanical strength but little elasticity. If a crack develops in a concrete block or concrete wall a built-up membrane with asphalt or pitch will have difficulty bridging the gap or resealing at underground temperatures. Despite some of these problems, built-up membranes have been used with success in earth-sheltered homes and other buildings.

Bentonite

Bentonite is a clay found readily in South Dakota, Wyoming, and Colorado. The development of bentonite embedded in cardboard panels is an effective waterproofing system since it expands greatly when coming in contact with water. The clay seals itself against further water penetration as it tries to expand. Bentonite has some resealing and bridging abilities. Since the material is natural it will not decompose with time.

Bentonize

This material mixes bentonite with a pastelike cement and is liberally sprayed to 3/8" thicknesses onto wall and roof surfaces. It has bridging and resealing abilities similar to bentonite. Leaks are localized since water does not travel easily beneath the membrane. The material is effective if it is uniformly sprayed to the required thickness.

Polyethylene Sheet

This inexpensive material is used widely beneath concrete floor slabs as a dampproofing product. It is a good barrier against capillary draw and vapor transmission. Polyethylene degrades in sunlight but will last a long time if completely covered underground. It has the ability to bridge cracks, but will not resist water pressure since laps are seldom sealed.

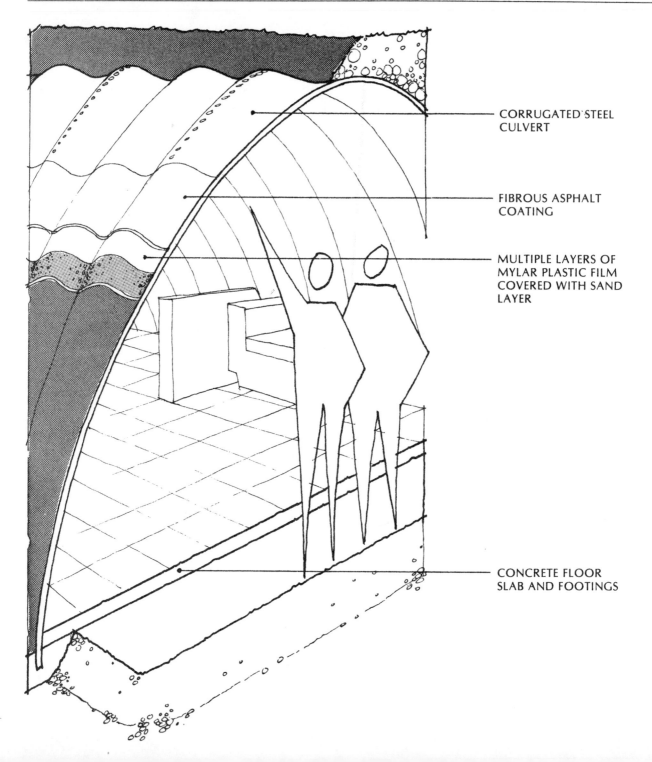

CORRUGATED STEEL CULVERT

FIBROUS ASPHALT COATING

MULTIPLE LAYERS OF MYLAR PLASTIC FILM COVERED WITH SAND LAYER

CONCRETE FLOOR SLAB AND FOOTINGS

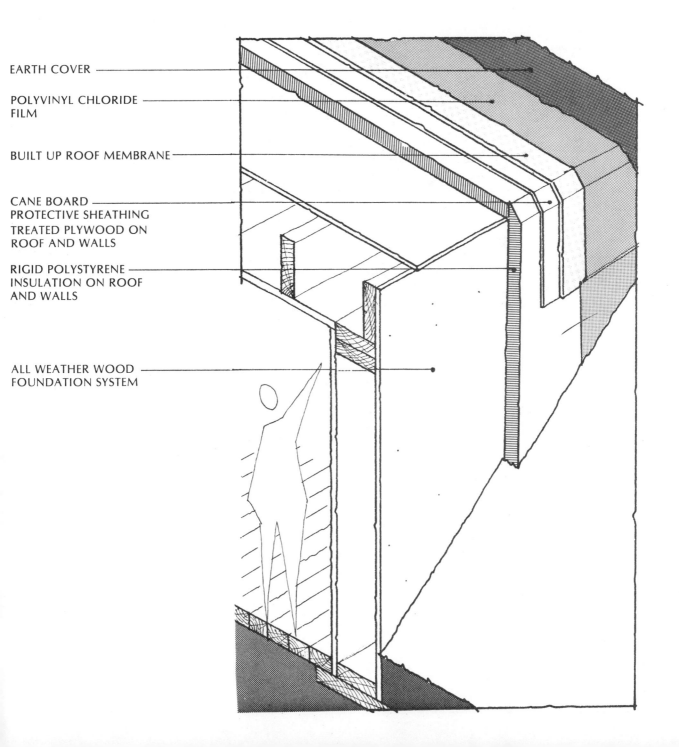

EARTH COVER

POLYVINYL CHLORIDE
FILM

BUILT UP ROOF MEMBRANE

CANE BOARD
PROTECTIVE SHEATHING
TREATED PLYWOOD ON
ROOF AND WALLS

RIGID POLYSTYRENE
INSULATION ON ROOF
AND WALLS

ALL WEATHER WOOD
FOUNDATION SYSTEM

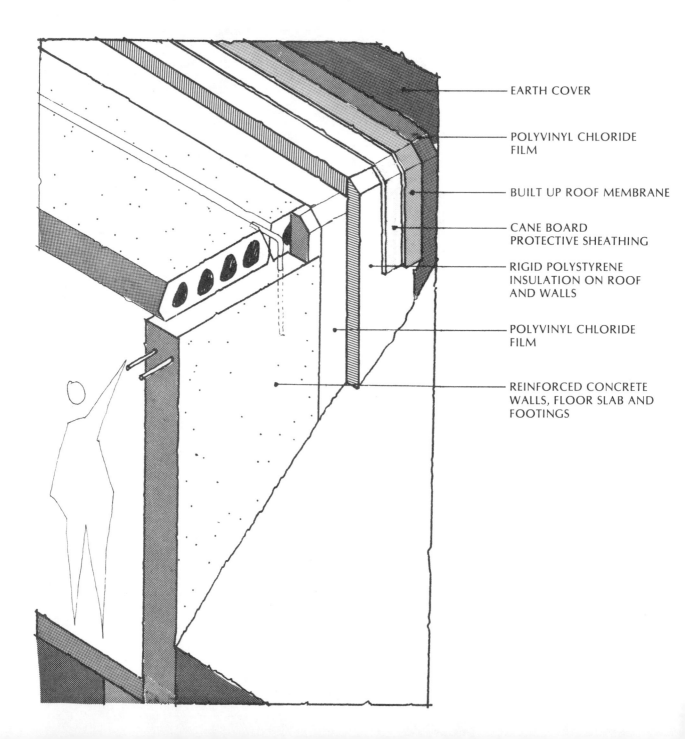

EARTH COVER

POLYVINYL CHLORIDE FILM

BUILT UP ROOF MEMBRANE

CANE BOARD PROTECTIVE SHEATHING

RIGID POLYSTYRENE INSULATION ON ROOF AND WALLS

POLYVINYL CHLORIDE FILM

REINFORCED CONCRETE WALLS, FLOOR SLAB AND FOOTINGS

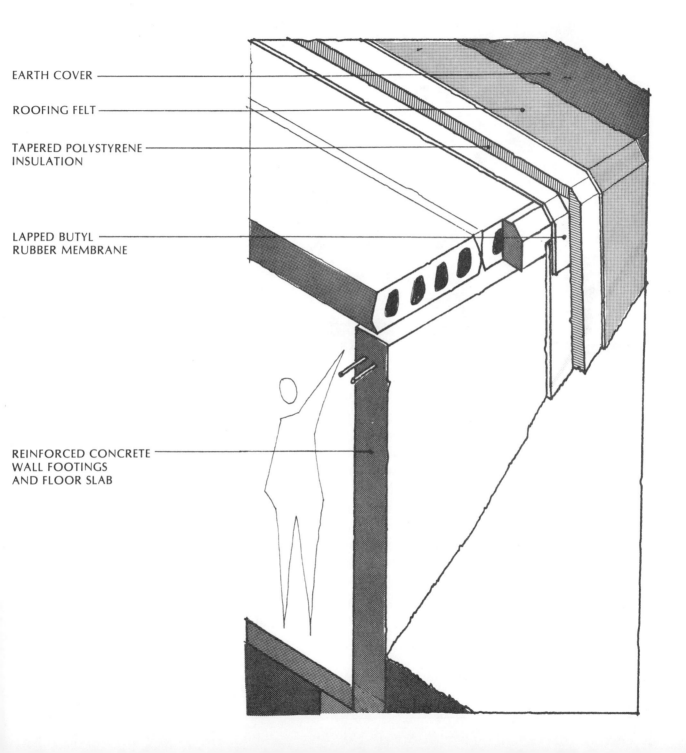

EARTH COVER

ROOFING FELT

TAPERED POLYSTYRENE
INSULATION

LAPPED BUTYL
RUBBER MEMBRANE

REINFORCED CONCRETE
WALL FOOTINGS
AND FLOOR SLAB

The shapes are Nature-made, the result of erosion by wind and water. Cracks and holes dot the surfaces, suggesting some elementary shelter. But perhaps even more than these, the finished silhouette of each peak spells out a house. To a man looking for a roof over his head, the rocks were a godsend; he was easily tempted to deepen the natural recesses and cut out, so they say, a niche for himself. Since the porous rock is no more difficult to cut than hardened cheese, a man can in no time and without much exertion excavate a good size apartment, including some stony furniture: tables, benches, and couches, not to mention fireplaces.

Cappadocian Cave Dwellings

Rudofsky, The Prodigious Builders (1977) p. 41

5 SUBSYSTEMS

Fuels

Heating Systems

Plumbing Systems

Electrical Systems

Subsystems are smaller elements of the home
which insure health, safety, and human comfort.
This chapter deals with the subsystems used in
conventional above grade homes. It describes
how conventional systems, with modifications, can
be adapted to earth-sheltered homes. Heating,
ventilating, and air conditioning systems condition
the interior spaces for the comfort of the
occupants. Water supply systems are essential for
human consumption, sanitation, and comfort. The
efficient disposal of fluid waste and organic matter
is critical in maintaining sanitary conditions within
the home and the surrounding site. Electrical
systems supply light, heat, and power to the home.

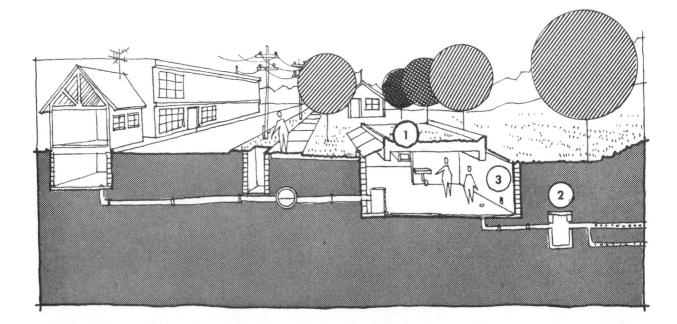

Fuels

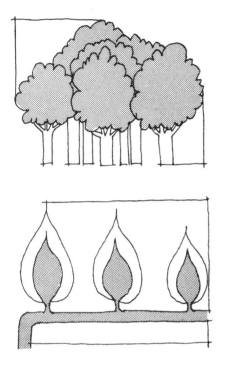

Today's modern technology has made available to us many fuel types including wood, natural gas, coal, oil, electricity, and solar power. The choice of which fuel to heat an earth-sheltered home with is dependent on three factors: cost, availability, and efficiency. In the past relatively inexpensive fossil fuels such as coal, oil, and natural gas made wood almost obsolete as a fuel source. It should be mentioned that wood should only be used if it is abundant without destroying existing trees. Active solar systems are still unable to compete cost wise with the fossil fuels. However, passive radiant energy from the sun is virtually free and should be exploited in the home.

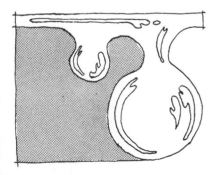

Availability is a determining factor in the choice of a fuel. Usually, when a fuel source is located a great distance from the home the overall cost will be high. Therefore, it is more practical to select a fuel located near the home. This will also help insure prompt delivery.

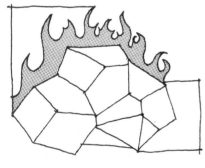

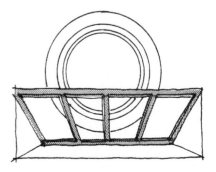

1 Heating Systems

Conventional home heating systems can be used in an earth shelter with minor adaptations. These include forced air, hot water, active solar, electric, and wood heating systems. The matrix (p. 187) graphically shows which heating systems can be employed in a concrete, heavy timber, A-WWF or steel culvert structural system.

UPFLOW

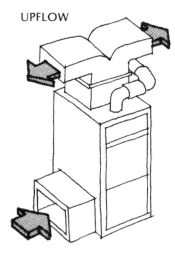

FORCED AIR

Forced air is the most commonly used heating system in conventional homes. In a forced air heating system air must be heated in a furnace and distributed in the home through a ducting network. All furnaces use the "up flow" or "down flow" principle. The up flow furnace requires supply ducts at the ceiling and return ducts at the floor. With a down flow furnace, supply ducts are placed in the floor and return ducts near the ceiling. The down flow furnace is recommended because it coincides with the natural air flow principle that hot air rises. Only the duct networks for down flow furnaces will now be discussed.

DOWNFLOW

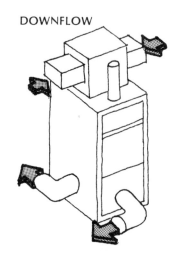

Diffusers are necessary for heat to be uniformly spread through the room. Baseboard and flush are the two types of diffusers available. The choice should be based on the appearance of the room

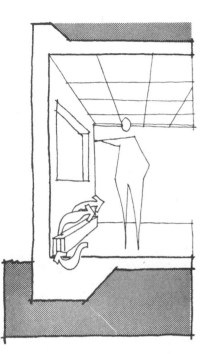

and the purpose of the diffuser. To reduce moisture buildup a floor to ceiling window or a sliding glass door should be heated with a flush type diffuser. For walls without windows a baseboard type diffuser is acceptable.

An advantage of forced air systems is that they control humidity, air motion, and purification. Important design considerations when installing this system in an earth-sheltered home are the location of the heating unit and of the ducting network.

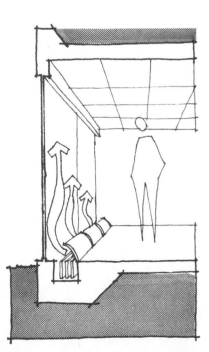

Locating the furnace near the center of the home will reduce duct lengths and heat loss. If the furnace room is not centrally located, efficiency will be lost.

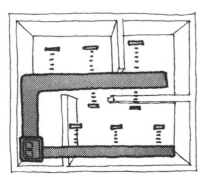
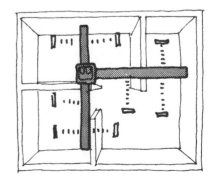

The furnace room can be placed against buried walls of the home since mechanical rooms do not require windows for emergency exit. This frees other areas for living space where windows are required.

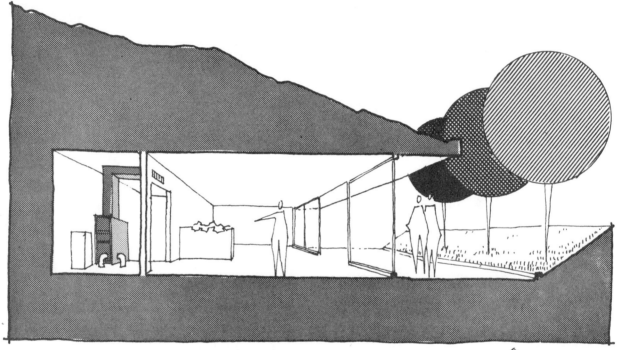

Concrete, wood, and steel culvert structural systems require different duct placement and installation techniques. Ducting can be buried in a concrete floor slab or placed above drop ceilings in a concrete structural system.

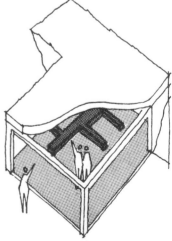

BURY DUCTS

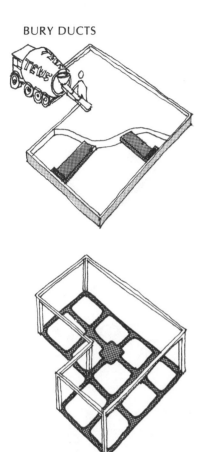

There are three basic layouts for burying supply ducts in concrete floor slabs. The three ducting layouts are the perimeter, radial, and loop systems. These systems must be positioned before pouring the concrete floor. Ductwork of the appropriate gauge must be selected to prevent collapse or buckling during the slab pouring process.

PERIMETER SYSTEM

Extensive ductwork is required for the perimeter system. Ducting for this system is laid out on a standard grid pattern. The diffusers are then located on the grid where needed. Warm air circulating through the ducting heats the floor slab. Heat loss is experienced until the floor slab is heated to the air temperature in the duct. This grid pattern insures even heat distribution throughout the floor. This system requires the excessive use of material and therefore can often be expensive.

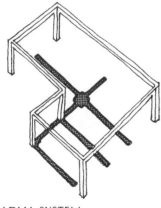

A radial system has minimal duct lengths and works most efficiently when the heating unit is centrally located. Armlike ducts supply warm air to each room offering individual thermostatic control.

RADIAL SYSTEM

The loop system layout is similar to the perimeter system layout. A main supply duct circulates air around the edge of the home. Feeder ducts supply air to the main duct, maintaining a uniform heat within the duct. This insures even and constant distribution of warm air to each room. Return air ducts can be concealed by a drop ceiling in a hallway or corridor. The ducts are suspended from concrete roof planks or rest directly on stud walls.

LOOP SYSTEM

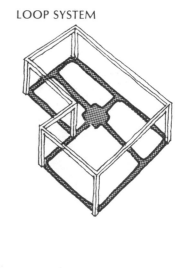

A-WWF and Heavy Timber Structural Systems

In a wood structural system, as with concrete, ducts can be run in the slab or along the ceiling. Wood systems also allow ducts to be placed in interior stud walls or in a crawl space under the floor.

CRAWL SPACE

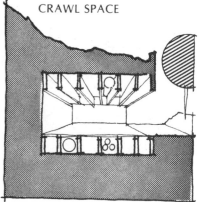

A pressurized floor system can be used as an alternative to ducts. It achieves air distribution by blowing warm air directly into a sealed crawl space. This allows air to enter the home through appropriately placed openings in the floor.

PRESSURIZED FLOOR SYSTEM

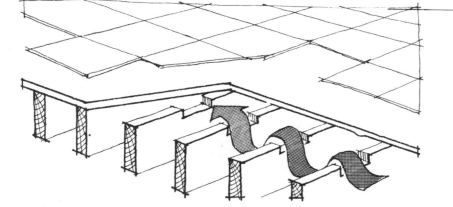

Ducting can be placed under a wooden floor in a crawl space. With the ducts located either between the joists or attached to the bottom edge, the layout patterns are identical to those discussed in the concrete section.

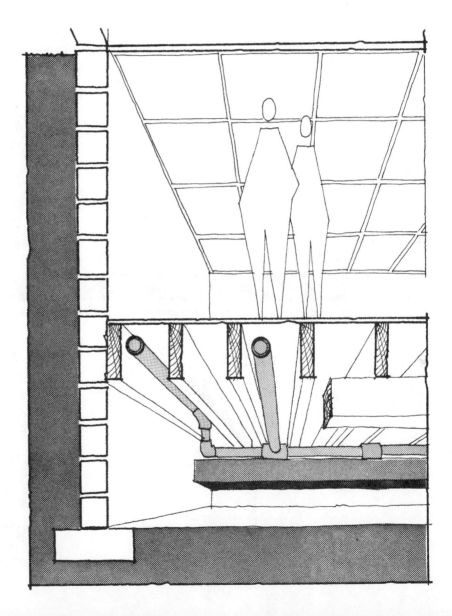

Interior stud walls are usually used for return air ducts. These ducts must be tied into a larger duct which runs along the ceiling.

In two-story construction, cold air return ducts can be buried in stud walls and suspended between or underneath second floor joists.

Steel Culvert Structural System

Ducting placement in a steel culvert structural system is similar to ducting placement in a concrete structural system. As an alternative to drop ceilings, ducts can be hung from the steel shell and left exposed. Both systems conceal ducts in concrete floor slabs or above drop ceilings.

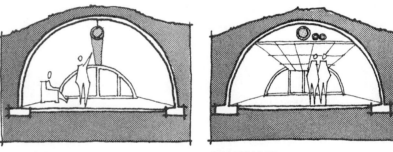

EXPOSED   CONCEALED

HOT WATER

A hot water home heating system has the same layout and placement implications for concrete, wood and steel structural systems. The water is heated by a boiler and then pumped through the home where convectors use the heat to warm the room. All hot water pipes must be laid out and positioned prior to pouring the concrete slab. Having the pipes embedded in the concrete slab helps eliminate wasted heat as the excess is trapped and stays within the floor system. The two types of water systems available are the one pipe and two pipe systems.

In the one pipe system a single pipe supplies and collects the water. Heated water flows continuously through a supply loop which runs around the perimeter of the home. The loop is tapped into by the convectors when heat is needed. Once the heat is drawn out of the water, the cooled water re-enters the supply line to return to the boiler and be reheated.

One continuous loop through the home is very inefficient. The convectors at the end of the loop may not receive sufficient heat. Heat is lost into the floor slab and decreases with each succeeding convector. Introducing more loops, possibly supplying different rooms, will insure sufficient heat where desired.

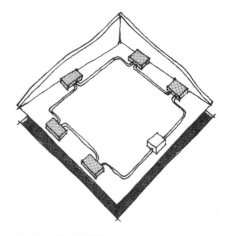

ONE PIPE SYSTEM

The two pipe system employs one pipe to supply water and one pipe to collect and return the water to the boiler. Heated water flows continuously through the supply line. Water is drawn off by the convectors when heat is needed. Once cooled, the water enters a separate collection pipe and returns to the boiler for reheating.

The two pipe system is more efficient than the one pipe system since cool water is not re-introduced into the supply line. This maintains a higher water temperature for heat use in the last convector in the line.

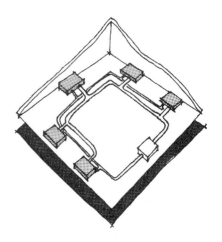

TWO PIPE SYSTEM

A convector can be positioned in the wall, on the wall or in the floor. This depends on window location and room layout. A floor convector is best utilized under a window which extends down to the floor.

This reduces moisture condensation on the glass. Placement of heating convectors on the wall allows for the use of fans which aid in the circulation of the air within the room.

Concrete floors and drop ceilings are commonly used for concealing ductwork. This makes the placement of ducting in a steel culvert structural system similar to that of a concrete structural system.

ELECTRIC HEAT

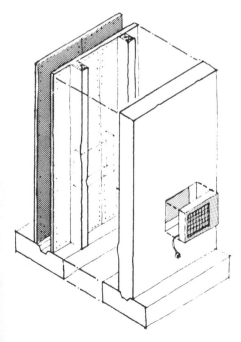

A radiant heat panel is a contained electric coil. It heats objects and masses within the room eventually heating the air. Radiant heat panels, like hot water heating systems, do not provide for humidity, air motion, or filtration of the air. To control humidity and filtration portable room humidifiers and filters must be added. Extra space is required for additional equipment.

Different radiant heating panel installation techniques are required for concrete, wood, and steel culvert structural systems. Additional formwork must be built to create an indentation for a unit recessed in a concrete wall. Electrical conduit must be placed prior to pouring of the concrete wall. Conventional framing is all that is needed when placing radiant panels in stud walls. In a steel culvert the radiant panels may remain exposed or be recessed in interior walls. (Angle brackets must be fastened to the steel culvert to hang the radiant panels.)

WOOD STOVES

Wood can be a practical fuel alternative if a large supply near the home is available. An immediate supply will lower or eliminate fuel costs. The earth around the shelter will help reduce temperature extremes. The combination of inexpensive fuel and earth sheltering insures a comparatively low heating bill. These facts make the wood burning stove competitive with other heating systems.

Wood burning stoves can be used as a primary or supplementary heat source. When used as a primary heat source a central location to maximize efficiency and provide even heat distribution is desired. For supplemental single room heating, the stove can be located in a corner or in the center of an inside wall. All combustible walls and floors require fireproofing. Fire resistant insulation, and asbestos wall board combined with a layer of floor brick provide the necessary fireproofing.

Open planning along with centrally locating the stoves will reduce the number of stoves required. Open square plans generally need one stove, whereas, a long rectangular plan may need more than one.

Concrete walls can increase the effectiveness of a wood burning stove. Placement near a wall provides an opportunity for the wall to collect and store heat. The wall will radiate heat to the remainder of the home.

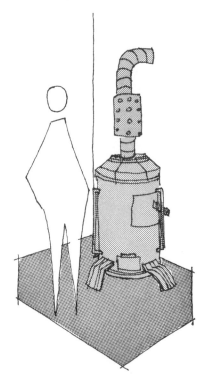

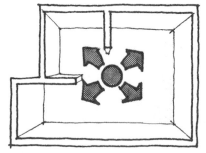

CENTRALIZED

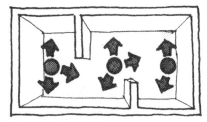

DECENTRALIZED

SOLAR HEATING

Energy from the sun, the potential of which is usually overlooked, can be harnessed to provide supplemental heat for a home. Passive and active solar systems convert the sun's energy into usable heat. Use of the sun's energy can potentially reduce organic fuel costs.

Passive Solar

Passive solar heat is produced by the penetration of the sun's rays through a home's roof, walls and especially window surfaces. Heat is collected, stored, and circulated without the use of mechanical devices. As an example, passive solar gain is experienced when a closed car is left out in direct sunlight and the interior heats up. This heat gain phenomenon is better known as the "greenhouse effect." The passive solar home must be both a collector and a heat storage container, acting as a "heat trap."

The home should be designed with solar collection in mind. The building form and placement of windows should maximize sun penetration during the winter and function as a sun screen during the hot summer months. The greater the amount of correctly oriented glass surface, the higher the potential for passive heat gain. (See sun section.)

The home must act as a heat sink or a storage container. Enough energy needs to be stored during sunny days by the home to provide radiant

heat during periods of no sunshine. An earth
shelter will function as an efficient heat sink if built
from adobe, concrete, or stone. These materials
are dense and are able to retain heat. The earth
around the home will contribute to the mass
needed to create a heat sink. Sunlight energy will
be absorbed by an object or wall when light falls
directly upon it. Sun should be allowed to strike as
much surface as possible to maximize this effect.
Water can also be used for heat storage since it
retains an enormous amount of heat and radiates
at a slow rate.

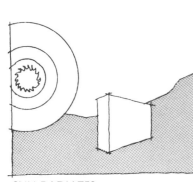
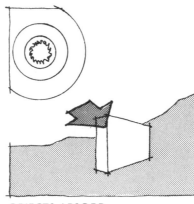

SUN RADIATES   OBJECTS ABSORB   OBJECT RADIATES

Heat that is gained should be trapped by the home
and not allowed to escape. Heat can escape a
home through walls, roofs, floors, windows, doors,
vents, and other openings in exterior surfaces. In
comparison to an above ground home, an earth
shelter is comparatively air tight. Cracks that allow
heat to escape are kept to a minimum. Covering
large glass areas with insulating blinds and shutters
reduces heat loss during periods of no sunshine.

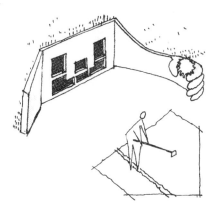
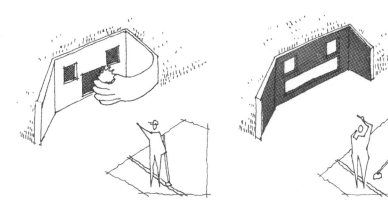

Active Solar

Collectors, transport networks, and storage facilities are the three main components of an active solar heating system which transforms the sun's energy into usable heat for a home. In cold climates it is not advisable to depend solely on an active solar system for all heating needs. Long periods of cloudiness and extreme cold will deplete stored heat, creating a need for a conventional backup heating system.

collectors

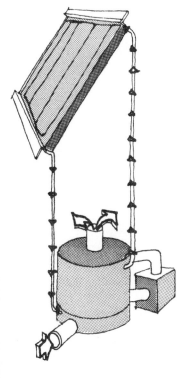

The solar collector is a self contained unit which basically consists of a glass cover and a heat absorbing surface enclosed in an insulated container. Sunlight passes through the glass cover trapping the sun's rays. The absorber, which is painted flat black, uses the sun's energy to heat water or air. Pipes or ducts, which are sandwiched between two layers of metal or welded to one face, carry the fluids which are heated by the sun's energy. This type of hot air or water solar collector functions with greater dependability and requires less maintenance due to enclosure of the air and water by ducts and pipes.

The open type of absorber allows the water to flow over the absorptive surface. Trays at the bottom of the collector gather the heated water to be pumped to storage tanks. There is a greater risk of malfunction with an exposed water flow collector. The absorber surface requires more maintenance than the absorber surface in a closed system.

Air collectors will not corrode. Cool air pumped into the low end of the collector gathers heat as it rises and passes over the warm absorber. Ducts are used to transport his hot air to storage beds.

For best results, the sun's rays should fall perpendicular to the collector surface. During the winter the sun is lower in the sky, so the collector should be positioned at a steeper angle to maximize absorption. Usually the collector is positioned at an angle 15 degrees greater than the local latitude.

Active solar collectors can be successfully incorporated into an earth-sheltered home. The installation of collectors on the flat roofs of earth-sheltered structures may present problems. Additional berming may be necessary to protect the backside of the collector from wind damage and heat loss. Collectors should not be mounted more than one high. Additional bracing as well as a stronger structural system may be required to carry the extra earth load and solar collectors.

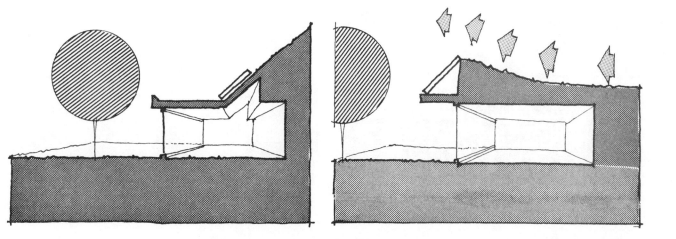

Collectors should not be positioned on an earth shelter where they will be shaded by nearby vegetation and buildings. If trees are located near the south facade, the collectors can be located to the rear of the roof for proper exposure. Collectors can be mounted vertically on wall surfaces which are free of windows, doors, or other openings. Vertical collectors are not as efficient in gathering heat as their tilted counterpart, but the added expense of framing may not justify tilting the collectors. Solar collectors do not necessarily have to be placed on the earth-sheltered home. Adjacent hills or berms can be altered for the placement of collectors. This eliminates the need for additional framing and structural support typical of rooftop mounting. However, if solar collectors are placed into hills or berms, extra insulated piping or ductwork is required to transport the heat to the home.

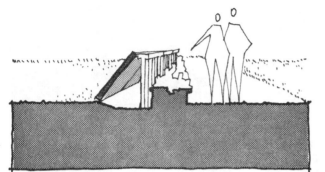

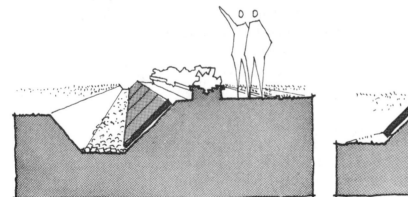

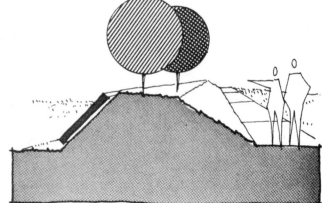

"Solaria," designed by Malcolm Wells (see structural systems), incorporates an efficient collector mount system. The sloped southern facade of the home makes mounting solar collectors easy since no additional structure or framing is required to tilt the collectors.

transport systems

Transport systems carry the heated water or air to storage tanks or beds. Water systems require piping while air systems require ducting. The piping and ductwork layouts are treated basically the same as hot water and forced air heating systems. Pipes, being smaller, require less space and can be run through concrete ceiling planks or through walls. Ducts require concealment above a dropped ceiling. An earth-sheltered home can be built with a warm air plenum enveloping the entire home. This plenum or air space is the space between interior stud walls and ceilings and exterior structural concrete or concrete block walls. Warm air circulates directly from the solar collectors into this space, eliminating the need for warm air ducts. Although this technique of creating an inner and outer wall is expensive, it can eliminate the need for heat storage facilities because the outer concrete wall acts as a heat sink.

Storage Facilities

Storage units are used to store extra heat not used by the home during the day. Three mediums are available to store heat generated by the sun's rays. These include water, crushed rock, and salt. The storage medium and type of collector must be chosen in combination with each other. A water collector always requires water as a storage medium. Air collectors require a medium consisting of small rocks or containers of water or salt.

Heat storage capacity is the amount of heat which can be stored in one cubic foot of material for one degree of temperature rise. Water is inexpensive, readily available, and has a high storage capacity. Water can be stored in steel tanks or concrete containers located in a mechanical room. Steel tanks are subject to the corrosive actions of chemicals in the water or antifreeze, while concrete is subject to cracking. Water contact and cracking problems may be alleviated by the use of plastic liners in the tanks.

Air collectors are used in conjunction with rock storage beds. Large rock beds, usually located beneath concrete floors or in crawl spaces, can be used to store heat for use during sunless days. Rock is less efficient in terms of heat storage than water. Therefore, a larger space is needed for a rock heat storage bed. Fans and ductwork are required to circulate air in and out of the rock bed.

Salts with a low melting point can be used to store heat from the sun. The salt is packed in thin containers which can be easily stacked in a small room or closet. Warm air from the collector is blown on and through these containers. As the warm air passes over the salt, the salt melts and heat is stored in the process. Heat is released when the salt cools and solidifies. Of the three mediums, salt can store the most heat per pound, but is also the most expensive.

2 Plumbing Systems

Water supply and waste treatment are important
subsystems in building design. When designed and
built properly they will insure a safe and healthy
environment. Plumbing systems include water
supply, distribution, and waste disposal.

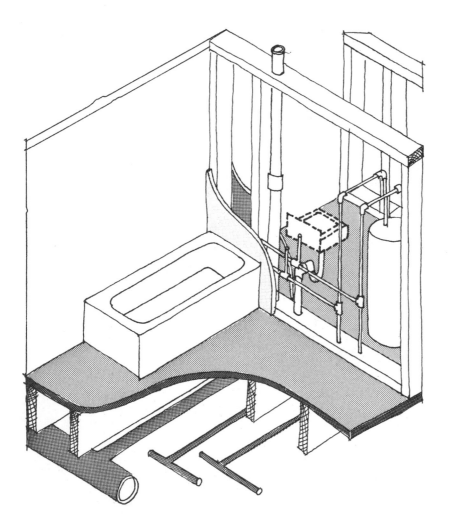

WATER SUPPLY

Water supply is obtained either from a public system when available, or from a private well when building beyond a public water system. A public water supply system offers a convenient and pretreated supply of fresh water. Pretreatment includes decontamination and softening. If a public water supply system is not available then a private well must be either dug, driven, or drilled. Decisions pertaining to location and type of well used, should only be made after consulting an expert. Driven wells are the most common today. Prior to buying a site, the depth of the water table should be determined if a well is needed. Higher development costs are encountered when a deep well is required because the cost of the well is based on its depth. By consulting neighbors during the dry season, information as to the depth of the water table in its worst condition, will be obtained.

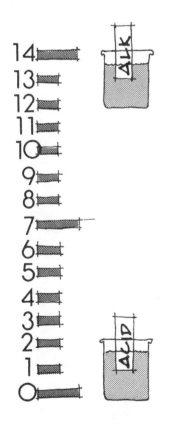

The predominant soil in our area is a sandy mixture, acidic in nature. Minerals contained in the soil are absorbed by water and can potentially cause problems in the home. Plumbing systems can corrode and become blocked. Hardness and discoloration can affect laundry and plumbing respectively. A scale measuring the acid and alkaline content of ground water is called the PH scale. The scale's numbers range from 0-14, 0 designating a very acid water condition. The neutral and most desirable point is 7. Minerals with a high acid content are mixed with alkaline chemicals to achieve a more neutral state.

A high acid content in water will cause iron pipes to corrode and eventually become blocked, thus restricting water flow within the home. Brass pipes will also be destroyed by this action. A device called a neutralizer can be used to raise the PH level to a safer and healthier level. Plastic pipe (PVC) can also be used as an alternative because it is not susceptible to the effects of acidic water. If treatment, other than the heating water is required, space must be allotted for the necessary equipment.

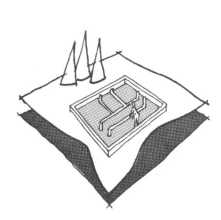

Distribution

A distribution system within the home carries water from the supply to the fixtures. Any water treatment required is usually done before distribution. Plumbing should be installed as straight as possible with minimal bends. Kitchens and bathrooms should be adjacent if possible to reduce the length and cost of piping. All necessary plumbing should be laid prior to pouring concrete floors in an earth-sheltered home. Burying all plumbing in concrete slabs is not advisable because maintenance is not possible. A crawl space allows all plumbing to be installed beneath the floor where maintenance is still possible. Additional fixtures can be easily added and serviced by a crawl space or access panel. Conventional plumbing techniques are applicable to earth-sheltered design when using standard stud wall construction. Piping can be run through the cavities in concrete ceiling planks. However, this complicates installation and makes repairs difficult if not impossible.

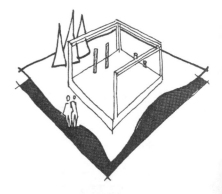

In the two-level earth-sheltered home the second floor system, usually of wood construction, can be used to carry and conceal plumbing. All plumbing carrying wastes must be kept as straight as possible with bends kept to a minimum. To prevent clogging all bends should be less than 45°.

WASTE DISPOSAL

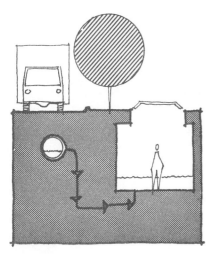

Building an earth-sheltered home where a public sewer system is available eliminates problems which might arise with a private septic system such as extra cost and construction delays. A waste disposal problem can develop in an earth-sheltered home when it is lower than the public sewer line. This situation requires the use of a sewage pump and collection tank usually located outside the home. The pump lifts the waste to a level where it can flow under gravity to the public sewer. The collection tank and inner pump are located adjacent to the home, the location of which must be considered during site planning.

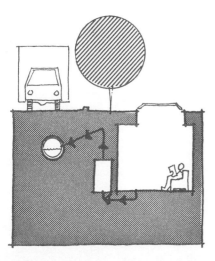

A private disposal system must be built when a public sewer is unavailable. The disposal system consists of a house sewer, septic tank, and waste distribution system. The house sewer, a cast iron pipe, runs from the home to the septic tank. Joints cast in concrete will insure tightness against leakage. As previously mentioned, soils with a high acid content will corrode cast iron pipes. To prevent breakdown the entire house sewer should be protected with a waterproof material to prevent contact with the acidic soil.

Septic tanks, usually concrete, hold waste material until bacteria can break down the solid particles into gas and liquid. The remaining sludge must be periodically removed from the tank. During decomposition, the liquid portion flows directly to the waste distribution system.

Absorption Test

A waste distribution system disperses liquid wastes from the septic tank evenly into the soil. The three types of waste distribution systems are drainage field, seepage pit, and sand filter. A major factor in determining the appropriate system for a site is the relative absorption of the soil. Relative absorption is the ability of the soil to accept water. This ability is determined through the use of a test pit dug on the site. The test pit procedure is as follows.

Begin by digging a hole twelve inches square by eighteen inches deep.

Fill the hole with water from a garden hose.

Allow the hole to drain.

Fill the hole again until the water is about six inches deep.

Time, 1" Drop In Minutes	Relative Absorption		
0-3	Rapid		
3-5	Medium		
5-30	Slow		
30-60	Semi-Impervious		
60+	Impervious		

Record the time it takes for the water to be completely absorbed and divide this time by six. The result is used with the table to the right to determine the relative absorption of the soil.

This absorption value in combination with the chart to the right reveals the possible alternatives. Example: a medium absorption rate will allow either a seepage pit or a drainage field to be used.

Relative Absorp.	Seepage Pit	Drain Field	Sand Filter
Rapid	●	●	
Medium	●	●	
Slow	●	●	
Semi- Imperv.		●	
Imper- vious			●

Two additional factors in determining a waste distribution system are the height of the water table and the amount of available land area. A drainage field can be used where the water table is high. A drainage field will not come in contact with a high water table because it is located just below the ground surface. Seepage pits, which produce high concentrations of waste, require that the water table be lower than 8 feet below the ground surface. Small sites usually dictate the use of a seepage pit, while a drainage field and a sand filter require more land area.

CASE STUDIES

In the selection of a waste distribution system soil type and slope steepness should be considered. These two conditions can be different for every site. Slope and soil combinations will be depicted in a fashion similar to the climatic information presented in the environmental impact section. The three generic soil types to be used are clay, sand, and gravel.

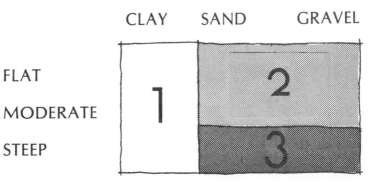

The three soil types will be analyzed in combination with flat, moderate, and steep slopes. The matrix shows nine combinations are possible with similarities among the solutions. The numbers indicate a grouping of the combinations which have the same solution.

1 Clay, All Sites

Clay has poor relative absorption and is a poor soil for waste disposal. Although it can be expensive, a sand filter disposal system can be used with a clay soil. The entire drainage bed must be excavated and filled with sand and gravel to allow for absorption. There are two types of sand filters, an open and a closed type.

The open type sand filter is located over a sand filled wooden trough above the ground. Liquid waste flows from the septic tanks through the drain tile above the sand. Holes in the drain pipe allow the liquid to drip slowly onto the sand. This technique is very unsanitary, creates undesirable odors, and is unacceptable.

The closed type sand filter is more acceptable since it is buried in a bed of sand. Liquid waste flows from the septic tank into the sand bed through gaps between the drain tile. These gaps should be covered with burlap or other porous material to prevent sand from clogging the tile. Collection tiles are placed below the distribution drain tiles to collect and carry excess liquids to more absorptive soil or a seepage pit.

2 Sand/Gravel, Flat/Moderate Sites

A sandy flat or sandy moderate sloping site is ideal
for a drainage field. Drainage fields are the most
desirable of the disposal systems since a more even
distribution of the waste water occurs. Sand has a
relatively high absorption capacity and will readily
accept waste. Sunny slopes, free of dense
vegetation, will increase the effectiveness of the
drainage field. The system consists of a distribution
box, drain tile and in some instances collection tile.
Waste water leaving the distribution box is
channeled to the various drain tiles.

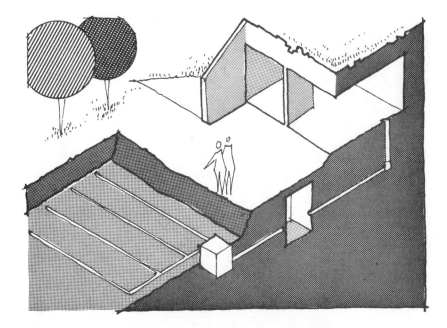

Tiles are placed in a trench two feet wide by two
feet deep on a bed of gravel, then covered with
fill. Slight gaps between the tiles allow for seepage
into the soil. A pitch of one inch drop per eight

feet of run is needed to maintain a constant and even flow of water through the drain tile. A slope of ten percent can accommodate a drainage field. The tiles must follow slope contours of a hill to allow the pitch requirement to be met with minimal problems.

3 Sand/Gravel, Steep Sites

A site is considered steep when a drainage field can no longer be used successfully. A seepage pit is used in this instance for waste distribution. A seepage pit is basically a holding tank for liquid

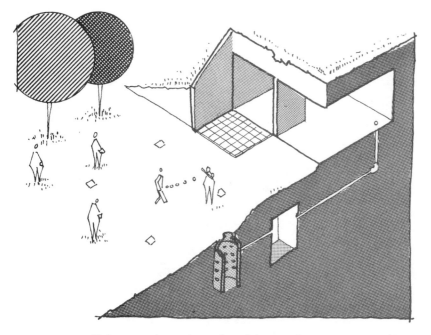

waste until it can be absorbed into the surrounding soil. This system is ideal for lots with a small area, steep slope and a low water table. The waste is deposited in a very concentrated area which would be harmful if a high water table is present. To maintain health standards a seepage pit should be located at least 100 feet from a well and 50 feet from a river or stream.

3 Electrical Systems

Electricity is used to power lights, appliances, and heating units within the home. An earth-sheltered home will have three basic circuits. They are general purpose, small appliance, and fixed appliance. Electricity is supplied to these circuits from an outside power source through a main breaker box.

General purpose circuits serve convenience outlets throughout the home except in the kitchen, laundry, and utility room. Many outlets can be placed on each circuit, because the expected power draw from lights, televisions, and stereos is low. If the home is small, each circuit usually supplies power to two or even three rooms. However, a hobby or workroom may need a separate circuit. Since occasional power tools might be used, more power may be drawn than anticipated.

Small appliance circuits, usually 120 volts, are used in the kitchen to supply power to counter top appliances. This circuit is similar to the general purpose circuit. However, fewer outlets are allowed on one circuit because of higher power demand of counter top appliances. Limiting the number of outlets on a circuit reduces the chance of a power overload.

SUPPLY

Electrical power is usually supplied to the home through an overhead power line. Burying this power line so it is unobtrusive is desirable in an earth-sheltered home. For ready accessibility in case maintenance is required and also to prevent accidental distrubance, a record should be kept showing where the cable is buried.

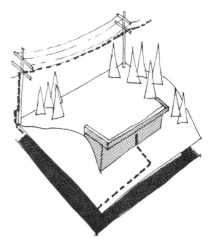

Heavy duty fixed appliance circuits supply power to the home's furnace, stove, and washer and dryer. Fixed appliance circuits require a heavier gauge wire. This is necessary to supply the 240 volts needed to power these larger appliances.

METER

Electrical power must pass through a meter prior to entering the earth-sheltered home. The meter monitors the amount of electricity used and must be located on the exterior of the home to be easily read by the electric company. Code requires that the meter be five and one half feet above grade to make it accessible.

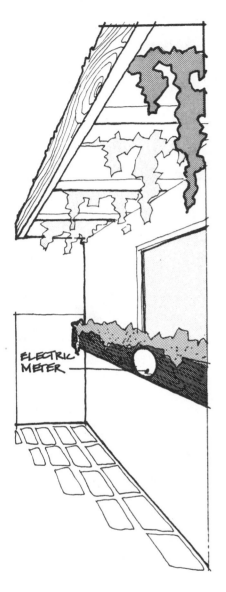

ELECTRIC
METER

Since earth-sheltered homes have few exposed facades, meter placement is usually restricted to the entry side of the home. The meter should be visible, yet visually integrated with the facade, so it does not become an eyesore.

WIRING AND CONNECTIONS

Installation of electrical wiring and power outlet boxes in earth-sheltered homes is similar to their above ground counterparts. Where no walls exist to conceal wiring, metal raceways or large conduit can be embedded in the concrete floor to carry wiring to desired locations. Plug in strips, an inexpensive alternative to outlet boxes, can be used to supply electrical power. Plug in strips can be easily installed above kitchen counters, can blend with floor moulding, and can be mounted on any existing wall surface. Standard metal hangers are used to attach conduit to concrete plank ceilings. These are similar to the hangers for heating and plumbing components.

Some ten million people live in the subsurface atrium dwellings of China's Loess Belt in the provinces of Honan, Shansi, Konsu, and Shensi. The inhabitants take advantage of the insulative qualities of the soil against the harsh winter winds of this region. The cave dwellings are easily excavated because the soil is unconsolidated. Most of these farmers live beneath their fields to make their homes cool in summer and comfortable in winter. The only hint of the presence of homes is smoke curling from small chimney shafts among the fields. Greater land productivity is achieved by placing homes below and fields above.

6 BUILDING CODES

The uniform building code is a set of minimal guidelines which need to be followed when constructing a dwelling. The codes were written for above grade dwellings and must be reinterpreted for an earth-shelter.

No uniform state codes exists yet, for one and two family residences, therefore local codes will govern an area or region. As a result, codes will vary from city to city. In addition, each inspector, being different, will further interpret the code and make the final decision.

Building codes have been divided into two categories. Codes with overall organizing implications and codes with detailing implications serve as examples of the many codes involved in building design.

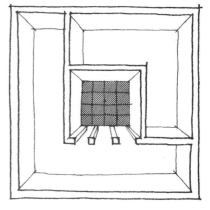

The City of Menominee, Wisconsin Building Code will serve as a model for reinterpretation of applicable codes pertaining to earth shelters.

Definitions

Several terms must be defined because of their important relation to earth shelters.

Courts: Open unoccupied spaces other than yards on the same lot with a building.

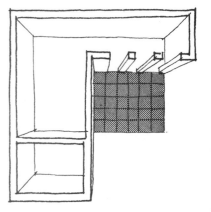

Inner Courts: Any court other than an outer court.

Outer Court: A court which extends directly to and opens on a street or other permanent open space, or on a required yard.

Grade, Building: This refers to the elevation of ground adjacent to the dwelling.

Habitable: A habitable room is any sleeping or living room.

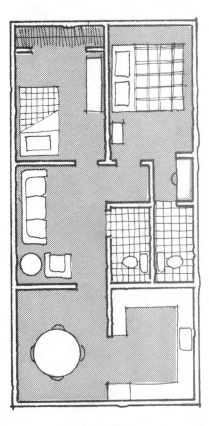

Living Room: A living room is one that is used for sitting, dining, or cooking purposes. This does not include: laundry, workshop, furnace, play or storage rooms.

Sleeping Room: A sleeping room is any room used for sleeping purposes.

ORGANIZING IMPLICATIONS

Exits   Section XIII

All first floor exits shall consist of doorways, platforms and steps with handrails and railings where required by code terminating at the outside grade.

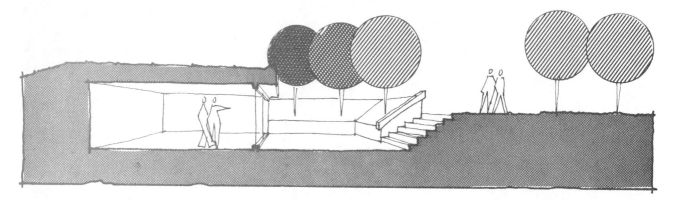

This implies that entries may be below grade if appropriate provisions are taken.

Each one, one and a half and two story dwellings shall have two exits.

Exits shall be arranged that in case of emergency when one exit is blocked, the other one is still accessible.

This implies that exits should be positioned as far apart as possible in an earth-sheltered home.

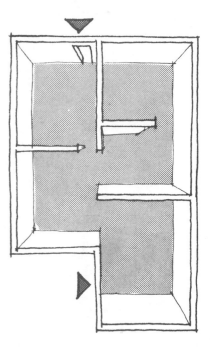

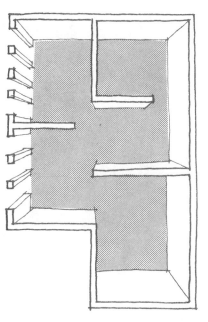

Windows   Section XXXI

The outer window or windows in every habitable room except toilet rooms shall have a total area of at least one-tenth of the floor area but not less than twelve square feet. The top of at least one such window shall be not less than six and one-half feet above the floor, and the upper half of it shall be made so as to open to the full width. The outside window in every toilet or bathroom shall have a total area of at least four square feet. Any toilet or bathroom must have a powerdriven fan, vented to direct passage of air to outside through roof or soffit vents if there is no window or skylight vent. This implies that every habitable room be located on the exposed side of the dwelling.

Courts   Section XXXI

Every room for human habitation, including bath and toilet rooms, shall be lighted directly upon a street or alley, or upon a court on the same lot with the building, or the sky. Every court, which is bounded on one side by a lot line, and which opens at one or both ends to the street, alley or yard, shall be an outer line court, and shall be at least three feet wide. Every court which is between two buildings or parts of buildings on the same lot, and which opens at one or both ends to the street, alley or yard, shall be an outer court and shall be at least six feet wide.

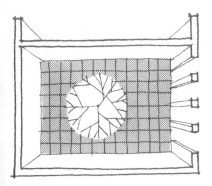

Every court which is bounded on one side and both ends by walls, and on the remaining side by a lot line, shall be an inner court and shall be at least ten feet wide and at least one hundred fifty square feet in area. No building shall be built, altered or enlarged so as to encroach upon the space reserved under this section for light and air.

Basement Rooms   Section XXXI

Every living and sleeping room in a basement shall be at least seven feet six inches high from the floor to ceiling, and ceiling shall be approximately four feet above the outside grade. The walls and floors shall be dampproofed and waterproofed. This implies that floor grade cannot be more than 3'-6" below existing grade. The other 4'-0" may be bermed.

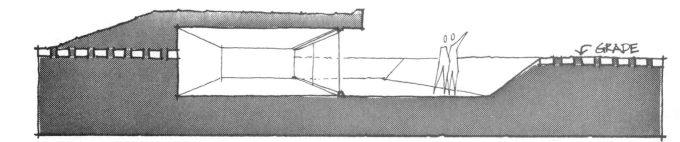

Stairs   Section XIV

All stairways and steps of more than three risers shall have one handrail located on the left hand side as you ascend and on the open side. Stairways and steps five feet or more in width shall have handrails on both sides. Handrails shall not be less than twenty-eight inches nor more than thirty two inches above the run of the stairs measured vertically from the top front edge of the tread to the top handrail. Stairway and step not required as part of the exits by this code shall have a uniform rise of not more than eight inches and a uniform tread of not less than nine inches.

Perhaps it is the fact that most caves are freely supplied by Nature which places them beneath contempt. Since caves are not offered for rent or sale, the idea of inhabiting them willingly rarely has occurred to humans. Despite the increasing unhealthiness of our surface life—the dangers compounded of poisonous air and polluted water, not to mention the everpresent dread of nuclear war—real estate agents have for the most part overlooked some startling opportunities. They go on peddling the brittle wooden crate, the plaything of floods and tornadoes, which promises no refuge from an angry environment.

Historically, from society's viewpoint, caves have always been considered for less than humans only; troglodytism—living in caves—amounted to disowning one's status as a human being. Still, we indeed know that caves have served as habitations much longer than ordinary houses; and however repugnant to us may be the thought of living in a naked cleft of rock, caves have been used repeatedly by man as a refuge from the ravages of the elements and as a safe place to retreat from his enemies.

7 AESTHETICS

Interior Space

The quality of the interior space is usually the primary consideration in the analysis of any home. Elements such as color, texture, material appropriates, natural or artificial light, acoustics, and the interplay of spaces can all make a home successful in terms of aesthetics.

In comparing earth shelters to above ground homes, in terms of aesthetics, a broader, expanded definition of aesthetics is required. Few of the five senses are generally used to appreciate an above ground home. Analysis seldom moves beyond the level of noting color and texture of interior/exterior finishes, the arrangement of rooms, and the number of closets in the house. Granted such information is important in any home, but for earth shelters to be appreciated aesthetically, they required a more indepth analysis. To coin an old cliche, "beauty is more than skin deep."

Photo by Darrow Watt

Photo by Creative Photographic Service

Photo by Darrow Watt

Photo and Design
by Coffee & Crier, Architects

An analysis of aesthetics of an earth-sheltered home should include how the home relates to the site, responds to climate, and in the efficiency and strength of its structural system. In addition, it is important to consider the feelings or moods generated by an architectural space. Earth sheltered homes convey a secure, almost womblike feeling. Quietness is one sense of security. They don't rattle when the wind blows or shake when a large truck or train passes. Interior living spaces can echo the caves our ancestors inhabited centuries ago or can be airy and open, alluding to more modern buildings which are products of recent technological advances. This quality allows earth-sheltered homes to respond to the different moods of people.

The structural or skeletal support system of an earth-sheltered home due to tremendous earth loads must be efficient and strong. Each structural system offers unique advantages and is usually selected in response to a specific site situation such as soil condition, and degree of slope. The steel arch structural system, with its curved walls, creates dynamic interior spaces. The interior textured surfaces of the steel arch diffuse light and absorb sound. Urethane foam, the primary insulating material in the steel arch, absorbs excess moisture from the air regulating humidity. This effect helps to maintain a more pleasant interior atmosphere. The brute quality of concrete, if properly exposed and patterned, can add aesthetic value to the interior of the home. An honest expression of the material can be achieved by illustrating its purpose. The concrete home can be shaped and sculpted into a true art form.

Photo by Al Drap

Exterior Space

Perspective by Malcolm Wells

Photo by
Creative Photographic
Service

Low profile earth sheltered homes don't attempt to dominate their sites but to blend naturally with them. These homes take advantage of their sites without destroying the qualities that made the sites desirable. They preserve natural amenities such as grass, shrubs, and other greenery. Earth sheltered homes can be equally complemented by either manicured lawns or wild ground cover.

Photo by Al Drap

Photo by Creative Photographic Service

Photo by Bill Heiting

Perspective by Malcolm Wells

Climate, an important design consideration, is carefully addressed by the earth-sheltered home. The sun for passive solar heating is usually maximized, whereas obstacles to wind, and water runoff patterns are minimized.

The earth-sheltered home shown utilizing the steel arch harmonizes with and complements the rolling terrain around it. The site has literally reclaimed this home by engulfing the roof-walls with vegetation. The ends of the culverts are, as one writer described, hooded eyes looking out to the world. Protruding chimneys, rock-like in character, blend into instead of disturbing the natural environment.

Elements of earth-sheltered homes, visible above the landscape, create interesting contrasts between manmade geometric forms and natural organic growth.

Photos by David Martindale

8 MATRIX

The matrices below begin to simplify the design process for the potential earth-sheltered home owner. The matrices combine and evaluate information in this book and make recommendations pertaining to the selection of structural, waste disposal, and heating systems. For example, given a soil and slope condition, the user can better select a structural and waste disposal system for his home. Once the structural system is chosen, an appropriate heating system can then be considered. The home owner will ultimately have to make the final selection of each system since all criteria for selection cannot be covered by this book.

STRUCTURAL SYSTEM

● MOST EFFICIENT   ◐ LESS EFFICIENT   ○ LEAST EFFICIENT

STEEP SLOPE TOP	CORRUGATED STEEL CULVERT	A-WWF	HEAVY TIMBER	REINFORCED CONCRETE
SILT	○	●	○	◐
SAND	◐	●	◐	○

MODERATE SLOPE	CORRUGATED STEEL CULVERT	A-WWF	HEAVY TIMBER	REINFORCED CONCRETE
SAND	●	○	◐	●
SILT	◐	○	○	●

FLAT SITE	A-WWF	HEAVY TIMBER	REINFORCED CONCRETE	CORRUGATED STEEL CULVERT
SAND	●	◐	○	◐
SILT		○	●	◐

STEEP SLOPE BOTTOM	A-WWF	HEAVY TIMBER	REINFORCED CONCRETE	CORRUGATED STEEL CULVERT
SAND		◐	●	◐
SILT		○	●	○

WASTE DISPOSAL SYSTEMS

SANDFILTER	CLAY	SAND	GRAVEL
FLAT SITE	●		
MODERATE SITE	●		
STEEP SLOPE	●		

DRAINAGE FIELD	CLAY	SAND	GRAVEL
FLAT SITE		●	●
MODERATE SLOPE		●	●
STEEP SLOPE			

SEEPAGE PIT	CLAY	SAND	GRAVEL
FLAT SITE			
MODERATE SLOPE			
STEEP SLOPE		●	●

HEATING SYSTEMS

	FORCED AIR	HOT WATER	ELECTRIC	WOOD STOVES	ACTIVE SOLAR
STEEL CULVERT	◐	◐	◐	●	
A-WWF	●	◐	◐	●	○
HEAVY TIMBER		●	●	●	◐
REINFORCED CONCRETE	●	◐	◐	●	●

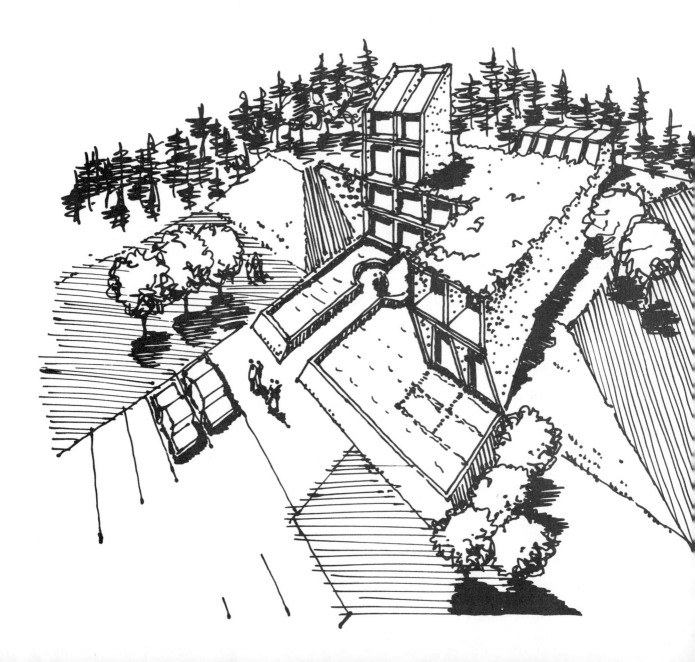

9 PROGRAMS & DESIGNS

Programming

This program focuses upon the site selection and design of a single family residence. Selection and planning of a site for an earth shelter is the first and most important aspect of the design process. For conventional above grade housing, site selection and planning is a routine procedure since energy efficiency is often overlooked. With any well designed energy efficient home it is important to understand the impact of the basic site. In most properly designed earth-sheltered homes the site protects its inhabitants as much as the building itself. On-site features include: orientation, location of adjacent homes, topography, soil type, vegetation and view.

Aside from the basic site, social factors are another set of criteria that must be evaluated. The distance to a job or shopping center often becomes critical. For clients with children, the distance to and from schools is another important parameter. These criteria as well as any others must be ranked from most important to least important when selecting a site. The ranking for every program is based on the needs and priorities of the client, therefore the ranking will be different in each program.

The clients for this program are primarily interested in an energy efficient retirement home. This is reflected in the ranking of site selection criteria.

1. Soil Type

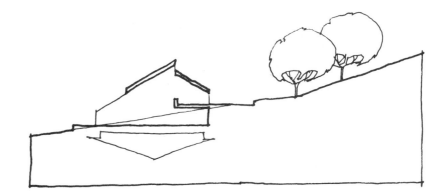

2. Orientation

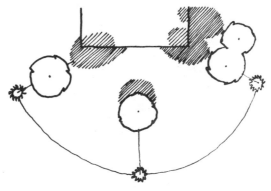

3. Social Factors
4. Topography

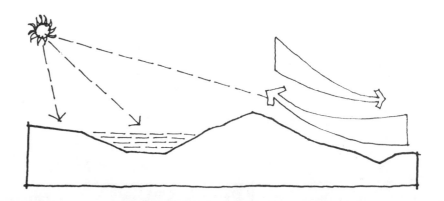

5. Adjacent Structures

6. View

7. Vegetation

The home should respond to the following specific room by room program requirements as well as these general characteristics:

Human scale
Earth look
Exposed structural system
Energy efficient features
Passive solar system
Informal atmosphere
Change of levels
Openess
Flowing spaces
Natural light
Minimal maintenance
Outdoor feeling inside

The budget for this project is around 80,000 dollars. The clients would like this home to have around 2,000-2,500 square feet.

FOYER

Activities

approach to home
greeting guests
removing of outer garments

Proximity

family room
to the street

Requirements

inviting
convenient to the street
storage of boots, umbrellas and wraps
weatherlocked if not located on the southern
facade

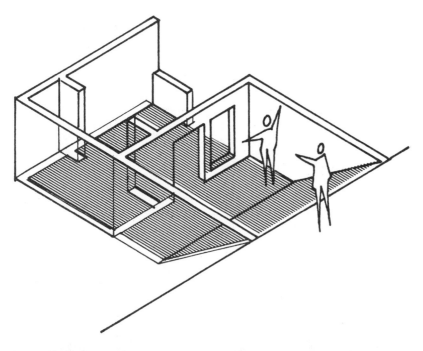

FAMILY ROOM

Activities

reading
talking or conversing
listening to music or the TV
entertaining guests
relaxing by the fireplace

Proximity

greenhouse
outdoor terrace or deck
dining room
recreation room

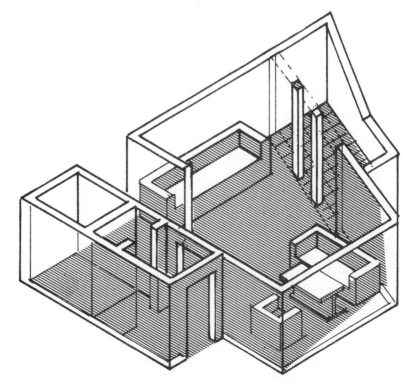

Requirements

The room must have a fireplace which may double function as a grill in the recreation room.
It must have a wet bar and storage for supplies.
Built-in shelves

Open to the greenhouse and out of doors.
Access to the greenhouse, outdoor terrace, and a view to the out of doors.

A relaxing, informal space for entertaining and casual relaxing, it must be large enough for twelve as might be the case of a party.

Skylight is desired whenever possible and artificial light will supplement.

SEWING ROOM

Activities

designing layout
cutting fitting
sewing

Proximity

laundry room
utility room

Requirements

Close to laundry room but not in it.
May double in the guest room.

This space may overlap physically with another space.

RECREATION ROOM

Activities

entertaining
games
parties

Proximity

bathroom
kitchen
family room

Requirements

The recreation room must have a wet bar,
pool table, benches or chairs, access to
the out of doors and storage.

Since the room will be used mainly in
the evening, skylight will not be needed.

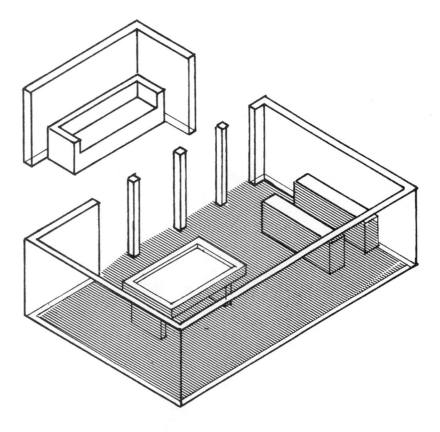

DINING ROOM

Requirements

The dining room must accommodate a five by seven foot table.
Storage and built in buffet.
Seat 12 persons.
The client would like variable lighting (artificial and natural) to create different moods.

This is the most formal room in the home. The space should have room for art display and china cabinet. It should be large enough for twelve yet intimate enough for two.

Activities

dining with guests
dining alone

Proximity

kitchen
greenhouse
family room

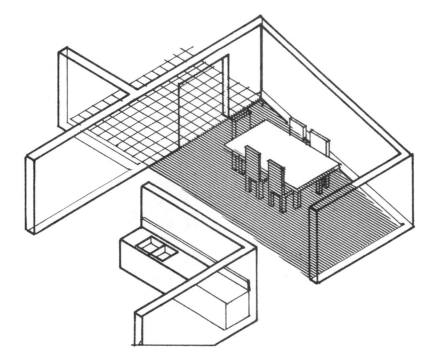

KITCHEN

Activities

food preparation and
storage
eating
writing

Proximities

dining room
family room
garage
recreation room

Intimate Dining

This in an intimate space
where one or two can eat
without feeling lost or in
the way. Dining, breakfasting
and snacking will take place
here. This area should not be
in the food preparation area
and would like to receive
some skylight.

Requirements

The kitchen should be integrated visually and/or
physically into the public space. Guests are welcome
to help prepare food according to the client. Since
much time is spent here it is important that the
kitchen be inviting and efficient.

Counter for eating with stools.

Wall pantry and closet.

Skylight with a viewing window.

The counter areas should be lit from above for
better visibility when preparing food.

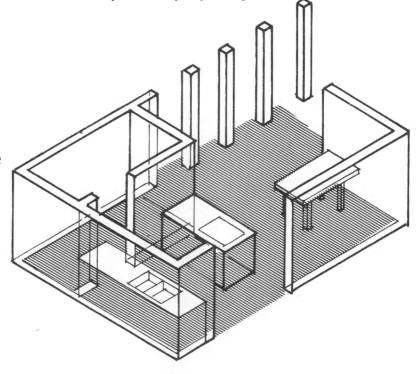

GREENHOUSE

Requirements

Must have a southern exposure
A healthy atmosphere for plants to thrive on

The greenhouse should be able to collect heat for the remainder of the structure during the winter months and expell excess heat during the summer months.

Access to the out of doors
Storage for gardening tools and supplies
Operable venting

Activities

potting
planting
sitting among
the plants

Proximity

family room
kitchen
dining room

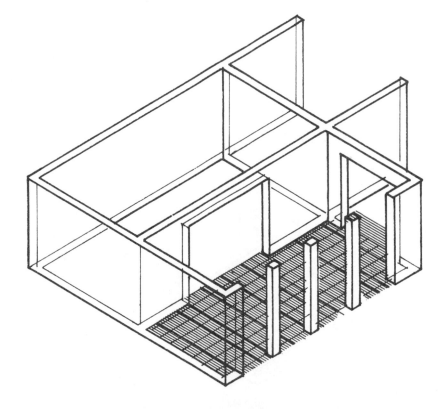

MASTER BEDROOM

Activities

sleeping
relaxing
dressing
reading
watching TV

Proximity

guest bedroom
sewing room
utility room
master bath

Requirements

The storage and dressers should be built in and adjacent to dressing room and bath.

Must be a quiet space
The bedroom shall accommodate a king size bed as well as two night stands.
Full walk in closet
The bedroom would enjoy eastern sunlight or skylight but not skylight in the late afternoon.

The bedroom should be a quiet place to relax and sleep. Preferably fill with morning skylight.
Room size is undetermined but should feel large.
A view is desirable.

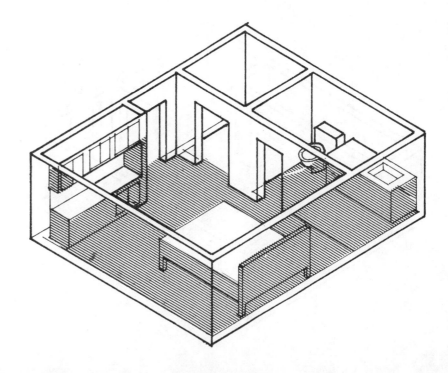

GUEST ROOM

Proximity

master bedroom
guest bathroom

Requirements

a relatively quiet, formal room with an
adjacent bath

provide storage for clothes

natural light and a view to the out of doors,
if possible

Activities

sleeping
relaxing
dressing

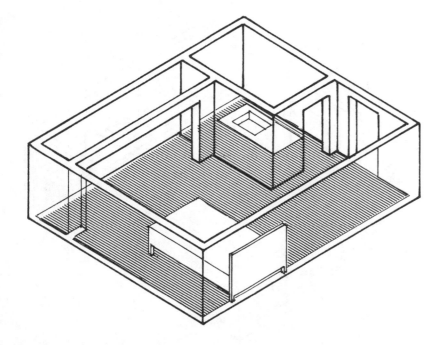

BATHROOMS

Activities

personal hygiene
bathing, relaxing

Proximity

master bedroom
guest room
public spaces such
as recreation room

General Requirements

full master bath
full guest bath
public half bath

Requirements for Master and Guest Bath

the bathroom fixtures should include a tub/
shower, vanity with mirror, double sink in the
master bath, and toilet

natural ventilation
natural ventilation is preferred as well as skylight over
artificial light

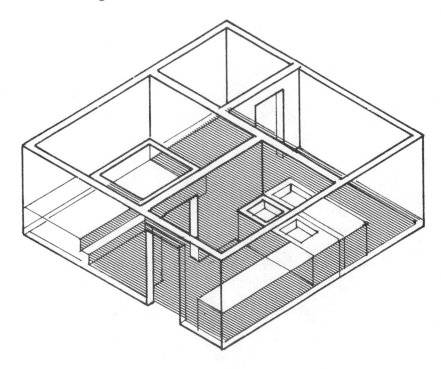

UTILITY/LAUNDRY ROOM

Requirements

The laundry room should be relatively close to the master bedroom so that dirty clothes don't have to be carried all over the house.

Must be vented to the out of doors.

This room must accommodate a washer, dryer, stationary tubs, stack space, ironing board, hamper and sorting space.

Activities

washing
drying
sorting
ironing

Proximity

sewing room
master bedroom

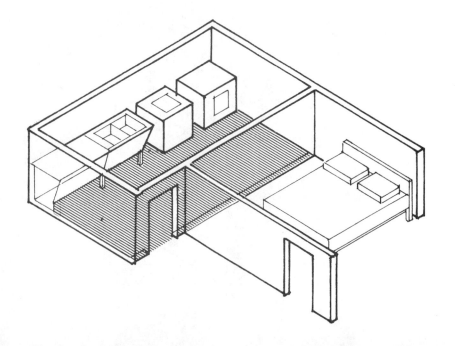

GARAGE

Activities

parking
maintenance
loading and unloading

Proximity

kitchen
workshop
family room

Requirements

Secure
Warm in the winter months so it may function as
an exterior workshop.
Built in cabinets for yard tools and maybe
lawn furniture.
Workbench for odd jobs and storage of
miscellaneous items.

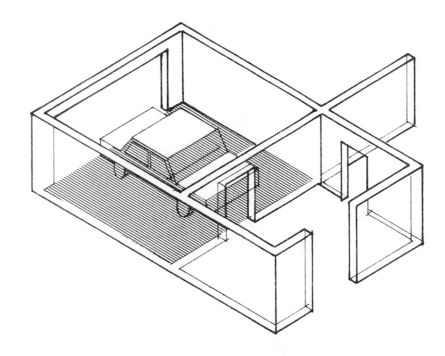

WORKSHOP

Activities

cabinet building
general repairs
odd jobs

Proximity

garage

Requirements

must be acoustically controlled
should have a direct connection to the garage to
bring in materials

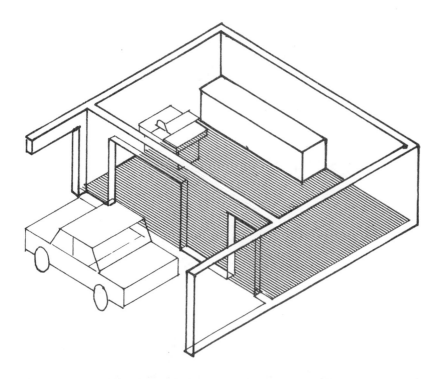

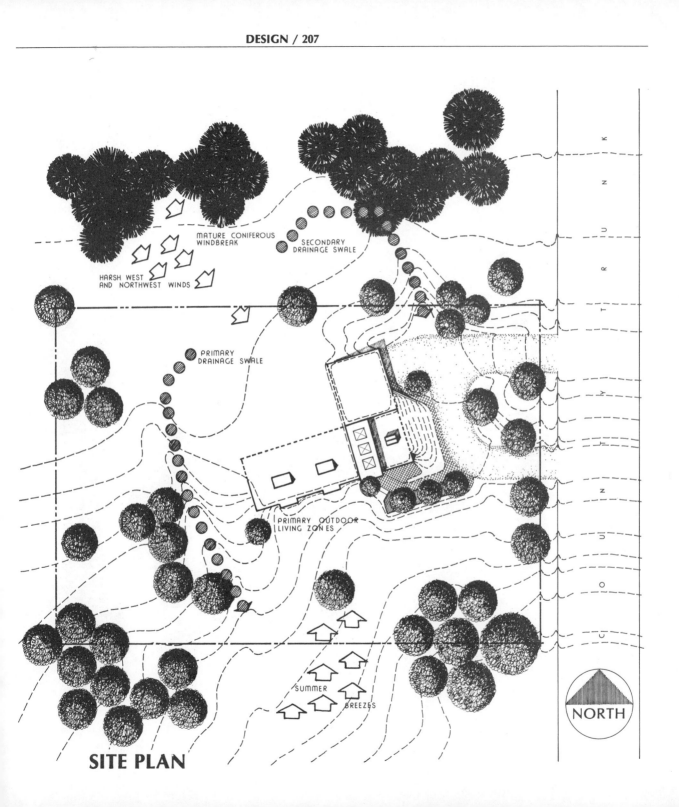

MATURE CONIFEROUS
WINDBREAK

SECONDARY
DRAINAGE SWALE

HARSH WEST
AND NORTHWEST WINDS

PRIMARY
DRAINAGE SWALE

PRIMARY OUTDOOR
LIVING ZONES

SUMMER
BREEZES

NORTH

SITE PLAN

perspective looking north

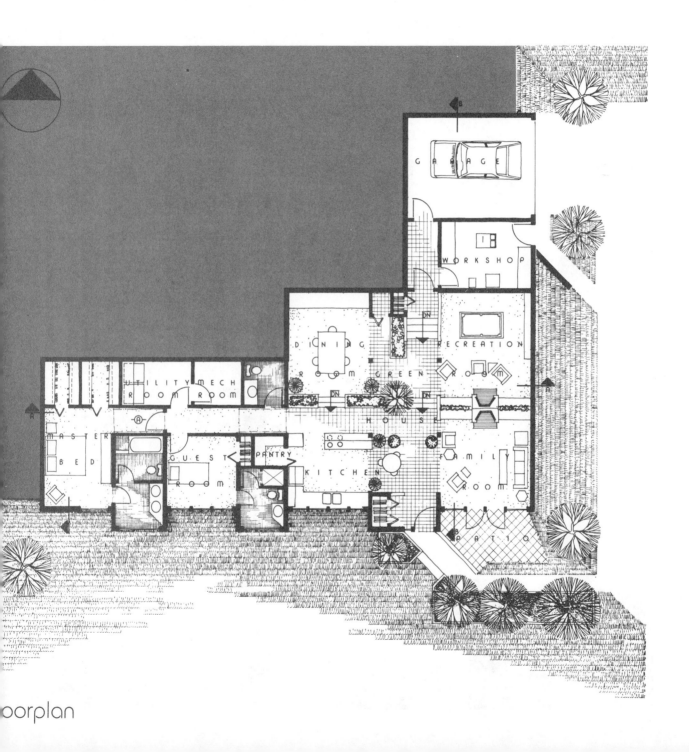

GARAGE

WORKSHOP

RECREATION

DINING
ROOM GREEN
ROOM

HOUSE

FAMILY
ROOM

UTILITY MECH
ROOM ROOM

MASTER
BED

GUEST
ROOM

PANTRY

KITCHEN

PATIO

oorplan

south elevation

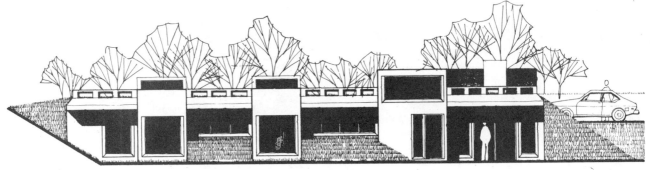

east elevation

section a-a

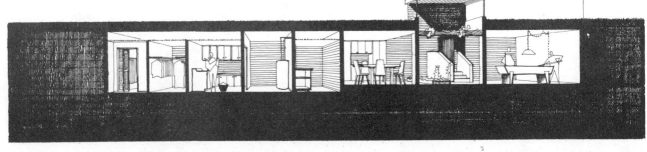

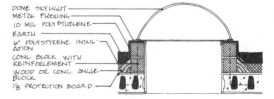

AXONOMETRIC AT GRADE

roof vent

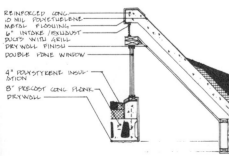

REINFORCED CONC.
10 MIL POLYETHELENE
METAL FLASHING
6" INTAKE /EXHAUST
DUCTS WITH GRILL
DRYWALL FINISH
DOUBLE PANE WINDOW

4" POLYSTYRENE INSUL·
ATION

8" PRECAST CONC PLANK

DRYWALL

skylight detail

DOME SKYLIGHT
METAL FLASHING
10 MIL POLYETHELENE
EARTH
6" POLYSTYRENE INSUL·
ATION
CONC. BLOCK WITH
REINFORCEMENT
WOOD OR CONC. ANGLE
BLOCK
⅛ PROTECTION BOARD

Programming

Centuries of seeking dwellings has taught man that those dwellings which respect the environment are the most successful. Earth-sheltered homes reflect a positive attitude toward, and concern for, living in harmony with the earth. An earth-sheltered home should be a true shelter as dwellings were originally intended to be. As a shelter the home should offer complete security and comfort to its inhabitants. The positive environmental effects, including cooling summer winds and warm winter sun, should be exploited by the home. Negative environmental effects, such as blistering summer sun and cold winter winds, should be minimized.

The earth shelter, instead of being merely an above ground home buried in the ground, should be of the ground, its forms being earth inspired. These forms should create a vocabulary which would be unique to earth-sheltered architecture. The earth shelter should function as a link or unifying element between the sun, earth, and man, thereby, enriching one's own awareness of himself, the environment, and the universe around him.

THE CLIENT

Clint Scherbarth of Rockford, Illinois is the prospective client for this earth-sheltered residence. His midwestern upbringing has given him a deep appreciation of nature and a desire to live in harmony with it. While in the Military, he traveled extensively in this country and abroad.

His experiences have given him a broad outlook on life and the problems being faced today. At the present time he is a sales representative for a Rockford electronics firm. He is currently single, although, he does anticipate a family life in the future. His interests include music, self-defense, traveling, cooking, camping and other outdoor sports. One of Mr. Scherbarth's major concerns is this country's ever increasing dependence on fossil fuels and their ever increasing costs. He is a steadfast believer in self-sufficiency and energy conservation. He hopes his home will be compatible with the goals of reducing energy consumption and costs.

THE HOME

Mr. Scherbarth would like his home to relate to both the past and the future. He would like his home to be primal and earthy relating it symbolically to man's past primitive dwellings. On the other hand, his work experience has given him an insight into the advantages of today's modern technology. He would like to see some uses of this advanced technology incorporated into his home as a way to grow towards the future.

The environment he feels a properly designed home can create, is a place where he can enjoy privacy and retreat from day to day pressures. He would like a home that would enhance rather than detract from the natural beauty of a site. Mr. Scherbarth believes many of the financial burdens of living today could be eliminated or reduced. He wants a home that is relatively maintenance free, with minimal energy consumption. An earth-sheltered home, Mr. Scherbarth feels, could satisfy many of these desires.

He requested that the home be basically one level, incorporating all the primary living spaces. He would accept a multi-level home if a better fit to the site could be achieved. A concrete structural system appealed to him because of the material's durability, capacity to store solar heat, and its ability to be decoratively patterned or textured. He preferred that materials for the home including concrete, wood, brick, native stone, glass, and ceramic tile be kept simple and earth tone in color. Mr. Scherbarth concurred with the notion of using minor floor and ceiling level changes to add interest and movement to the interior of the home.

The client requested the home be designed as a whole and then constructed in three phases. This would allow the essentials to be built immediately during the first phase followed by the remaining two which would be built in the future as additional financial resources became available. The development of a total master plan for the home would insure cohesive organic growth instead of sporadic confused expansion. This master plan should be considered flexible since his future needs could potentially change.

PHASE ONE

ENTRIES

Activities

Entering, leaving the home, putting on and removing coats, shoes

Proximity

Living room, kitchen

Requirements

Entries should have adjacent storage space, be weather locked, and provide a pleasant transition from exterior to interior space.

Entries should be daylit.

LIVING ROOM

Activities

Family gatherings
Listening to music, watching television
Playing cards, reading, sewing

Proximity

Dining room, kitchen, entry, greenhouse

Requirements

Living room should be a large and open space, accommodating family gatherings, yet be small and intimate.

Living room should be daylit, bright, and airy. Artificial illumination should be provided by room lamps and recessed ceiling spot lights.

A large fireplace, of masonry, should be provided, with heatilator, attached wood burning stove, and log storage area. Log storage area should be refilled through closeable opening in roof.

Floor should be heat absorbing concrete slab covered with earth-tone ceramic tile. Area rugs as opposed to wall to wall carpeting should cover the tile.

GREENHOUSE

Activities

Gardening, interacting with nature
relaxing

Proximity

Living and dining room, kitchen
Outdoor deck or terrace

Requirements

The greenhouse should be a strong visual attraction in the home. The character of the space should set the theme for the remainder of the home.

The greenhouse should be a dryspace and highly usable during winter months. During overcast or inclement weather it should be separable from the rest of the house.

A floor drain, spigot, and warm air circulation system to the home interior were requrested by the client.

KITCHEN

Activities

Cooking, eating

Proximity

Dining room, living room, laundry room, mechanical room, garage, entry

Requirements

The kitchen should be the functioning center of the home. It should be an informal space with pots, pans, cooking utensils, and food storage containers openly displayed.

The kitchen should be open, airy, daylit, and easily accessible from living, dining, and greenhouse areas.

If possible the living room fireplace should extend into the kitchen so an open grill can be provided. This grill should have a slide over counter when not in use.

There should be enough space for a double sink, refrigerator, built-in oven, microwave oven, countertop island, chopping block, and counters for food preparation.

All counters should be directly lit from above for better visibility during food preparation.

DINING ROOM

Activities

Eating, entertainment, family gatherings

Proximity

Kitchen, living room

Requirements

The dining room should be an informal space that can be used for breakfast, lunch, and supper.

The dining room should be large enough to accommodate a table that can seat up to eight people at one time.

Outdoor access from the dining room to a terrace or deck should be provided.

The space should be spotlit for different moods.

MASTER BEDROOM, BATH

Activities

Sleeping, reading, relaxing, dressing.
The client also wished to see the stars while lying in bed.

Proximity

Bedroom #2, greenhouse, outdoor terrace or deck.

Requirements

Bedroom must accommodate a king-size bed and several chairs.

Storage space should be built-in and be adjacent to bath and dressing space.

Bedroom should have general illumination with supplemental spot lighting. This room should be well daylit.

The client requested the bath be compartmentalized so it can be shared with bedroom #2.

The bathroom should have steam room potential with a wood slat floor. In addition the tub and shower should be of tiled lined concrete, separated from the master bedroom by a fixed glass partition so spatial contact is still possible.

Bathroom fixtures should include tub, shower, vanity with mirror, toilet, and storage.

BEDROOM #2

Activities

Sleeping, play and other activities of children, studying

Proximity

Master bedroom, bath

Requirements

This bedroom should be large enough for two children, including two beds, storage, shelves, and study space.

The room should have general illumination and be well daylit.

BEDROOM #3

Activities

Sleeping, studying, working, reading

Proximity

No requirements

Requirements

This room should be able to function as a bedroom, study, library, or den, or playroom.

The room should accommodate a large desk, have generous shelves and storage, and if possible overlook the living room or greenhouse.

The client requested that space be provided for a large aviary, aquarium, and terrarium.

GARAGE

Activities

Parking, washing and repairing automobiles, various shop functions.

Proximities

Street, kitchen, utility and mechanical rooms, wood storage area.

Requirements

Space for two automobiles, wall storage, or adjacent storage room should be provided. The storage area should have a work bench.

The wood storage area should be covered and have easy access by truck for simplified unloading of wood.

The client requested a floor drain with a metal grate, dirt trap, and a sloped concrete floor for better drainage.

A space should be provided for the collection and separation of refuse generated by the home.

MECHANICAL AND UTILITY ROOM

Activities

Laundry, maintenance on mechanical equipment.

Proximity

Kitchen, garage

Requirements

The mechanical room should have an emergency lighting system in the event of a power failure.

This room should be large enough to accommodate heating, water, and electrical equipment. Storage space should be provided.

The utility room should have space for a washer and dryer and a small sink. Storage space should also be provided.

ADDITIONAL REQUIREMENTS

The client requested that a maximum use of passive solar heating techniques be made. Instead of mechanical air conditioning, a natural ventilation system should be used. The home should be heated by a gas fired hot water system. Heat distribution will be through hot water pipes embedded in the concrete floor slab and hydronic wall panels. Wood burning will also be used.

An intercom system should be installed in the home, linking all rooms with the kitchen.

The client owns a large stereo system and requested that the components be built into the home. Speakers should be installed in the kitchen, dining room, and bedrooms, and linked to the stereo.

PHASE TWO

Space should be allocated in the living room during Phase One for the installation of a home computer.

An active solar heating system utilizing hot water should be installed on the roof. The system should be adjacent to the mechanical room and be sheltered from harsh winter winds. The computer mentioned above should be near the mechanical room to monitor the heating system.

The client has an interest in Astronomy and requested that a space be developed on the roof of the home to mount a telescope and observe the stars. This space or deck should be large enough to accommodate several people. If possible this deck should be accessible from the living room or master bedroom.

PHASE THREE

An area on the site should be selected for the erection of at least one windmill to generate electric power for the home. The windmill should be placed on the site to maximize wind potential.

The client asked that several shops be added near the garage for various hobbies and home maintenance activities. These include gardening, wood working, electronics, and auto repairing. These shops should be provided with water, heat, and electric power.

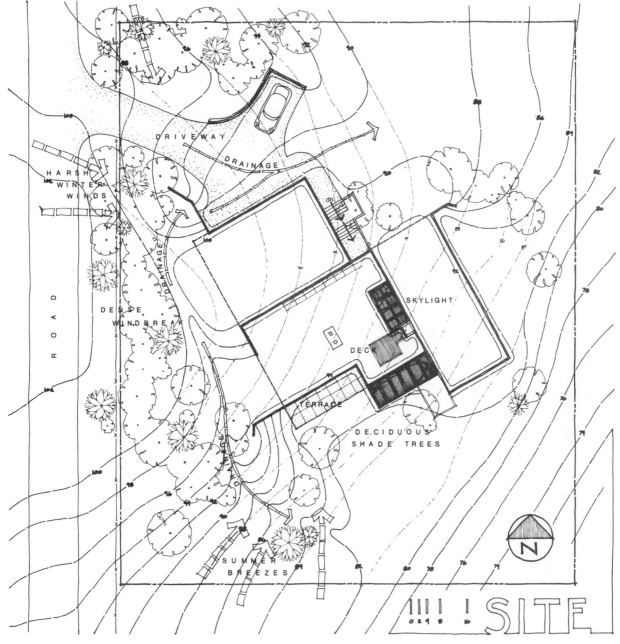

DRIVEWAY

DRAINAGE

HARSH
WINTER
WINDS

DRAINAGE

R O A D

DENSE
WINDBREAK

SKYLIGHT

DECK

TERRACE

DECIDUOUS
SHADE TREES

SUMMER
BREEZES

N

SITE

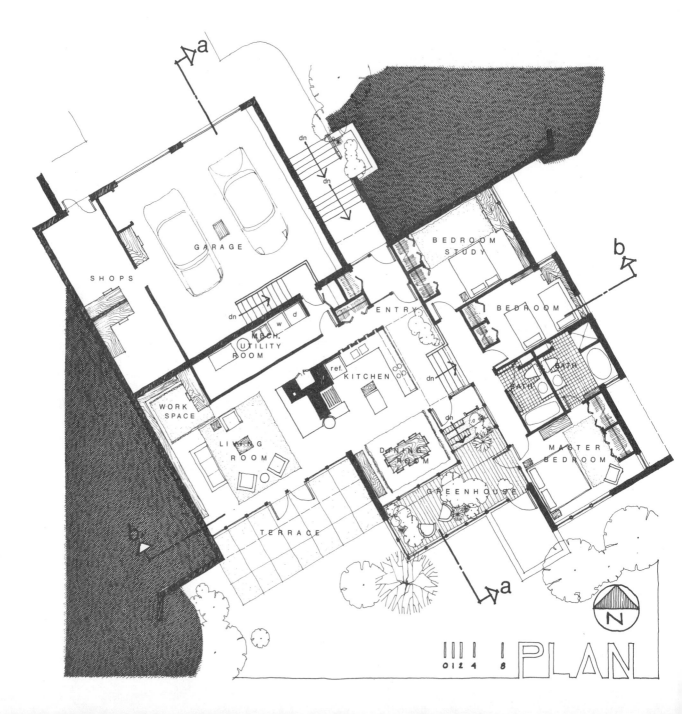

a

dn
dn

GARAGE

BEDROOM
STUDY

b

SHOPS

BEDROOM

dn   MECH.
UTILITY
ROOM
w   d

ENTRY

ref.   KITCHEN

dn

BATH   BATH

WORK
SPACE

dn

LIVING
ROOM

DINING
ROOM

MASTER
BEDROOM

GREENHOUSE

b

TERRACE

a

N

|||| |
0 1 2 4   8   PLAN

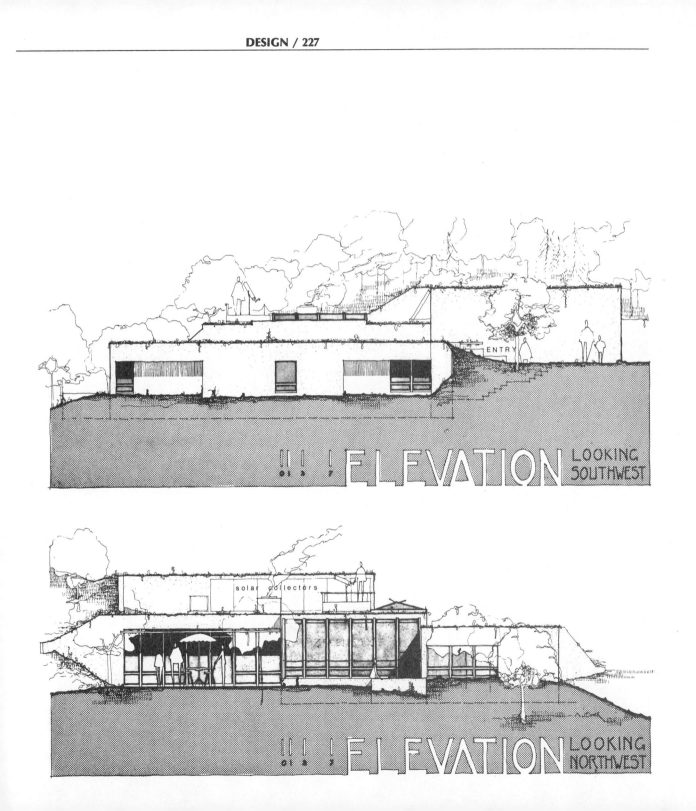

ELEVATION LOOKING SOUTHWEST

ELEVATION LOOKING NORTHWEST

a-a

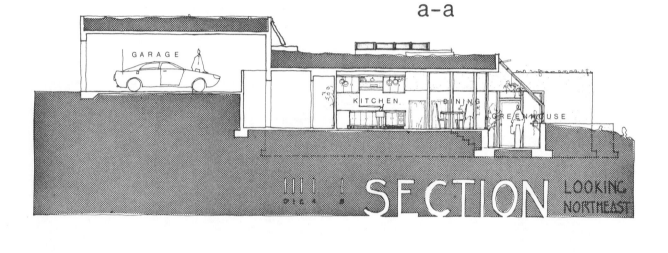

SECTION LOOKING NORTHEAST

b-b

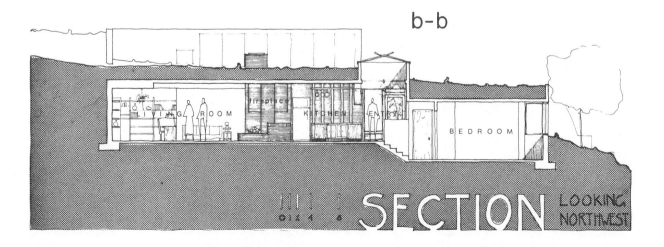

SECTION LOOKING NORTHWEST

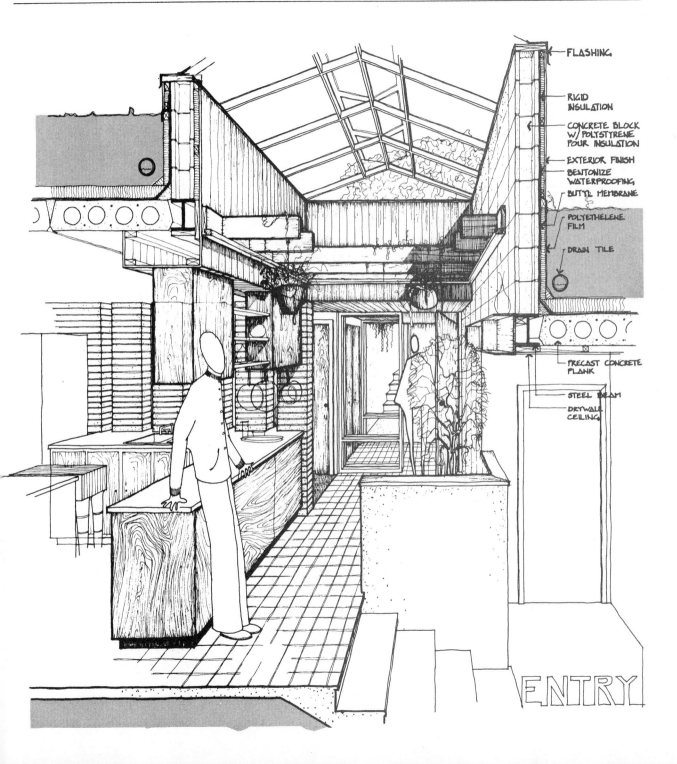

FLASHING

RIGID INSULATION

CONCRETE BLOCK W/ POLYSTYRENE POUR INSULATION

EXTERIOR FINISH

BENTONIZE WATERPROOFING

BUTYL MEMBRANE

POLYETHELENE FILM

DRAIN TILE

PRECAST CONCRETE PLANK

STEEL BEAM

DRYWALL CEILING

ENTRY

Programming

The following program focuses on the design of a single family shelter for a client of middle income. The clients Jim & Delores, expressed their interest in an earth sheltered home for two main reasons: preservation and protection. A favorite pastime of the clients' is gardening and landscaping. Earth sheltering allows many opportunities for this activity as well as preserving the landscape. Granted a home does not occupy much space, but the client's wish their shelter to become secondary to the elements of nature. It is desired to integrate the enclosed living environment as much as possible with the open natural environment. Energy conservation, and ultimately dollar conservation, is a concern of the clients. Earth sheltering is seen as a way to protect the clients' living environment from the cold of the Wisconsin winters, and the heat of the summer sun. Earth sheltering is a way of "turning your back" on the frigid winter winds, potentially reducing fuel costs. A site is desired where protection from the northerly winds as well as exposure to southerly sun can be accomplished.

When Jim comes home from work in the evening, he likes to get comfortable and enjoys reading and watching television. Delores, a homemaker, spends much of her time at home. She enjoys baking, sewing, and gardening. Periodically, Delores has a group of friends over for coffee and conversation. The clients request a modest shelter of about 2000 square feet, with a few characteristics as follows:

- integration with nature
- minimal maintenance to shelter exterior
- sandy cream brick exterior
- "L" shaped plan if possible to separate activivities
- open, informal
- sun room, greenhouse, or "front porch"

The clients set a budget of $60,000 for the project and wish to reside here after retirement.

LIVING ROOM

Activities

Entertaining guests, knitting, reading, conversation, relax by the fireplace, listening to music from the stereo.

Proximities

Dining room separated by a half wall, near the front entry, kitchen.

Requirements

A lannonstone fireplace to be the center of attention, a cathedral ceiling with exposed beams if possible, daylit, cozy, and airy.

KITCHEN/DINETTE

Activities

Preparing and cooking food, storage of food and cooking equipment, eating in dinette, coffee club meetings.

Proximities

Living room, laundry room, garage and secondary exit, half bath.

Requirements

Dinette to serve up to 10 at once, with the ability to overflow into the kitchen if needed. Plenty of counter space for preparation of food and plenty of storage for utensils should be provided. It is desired to have indirect lighting under upper cupboards for preparation work.

BATHROOM

Activities

Shower, shave, and dress.

Proximities

A full bath is needed near the bedrooms, and a half bath is needed near the kitchen.

Requirements

Double sink in the full bath, storage for linen, toiletries, etc.

ENTRY

Activities

Hang coats, store boots, shoes, etc., provide a weather lock during the cold months

Proximities

Living room, kitchen, garage

Requirements

Must have double door air lock for protection. Must be inviting, from outside and within. "Front porch" use entry as a green transition of outside to inside. The front porch can be used as a seating area too.

FRONT PORCH/SUN ROOM

Activities

Grow plants, enjoy the sun, conversation, mix with nature.

Proximities

Garage, front entry or secondary entry.

Requirements

Provide a space for growing plants with a seating area, like an indoor garden.

MASTER BEDROOM

Activities

Sleeping, dressing, reading.

Proximities

Near to the bathroom. It is desired that the bedroom be away from the main living spaces.

Requirements

The space must be quiet with acoustical separation among rooms. Plenty of closet space should be provided. The bedroom should be oriented toward the morning sun so it will wake the clients up. The room should evoke a cozy, relaxed feeling. A sink should be provided for shaving and makeup application.

BEDROOM #2

Activities

Sleep, dress, listen to music, practice writing skills, assembling puzzles and drawing.

Proximities

Master bedroom, bathroom.

Requirements

Plenty of closet space, room for a desk, chair and/or table.

BEDROOM #3

Activities

Temporary sewing room, den, etc.

Proximities

Master bedroom, bathroom.

Requirements

Temporary storage of out of season clothing, space for a hide-a-bed, sewing machine, and table.

LAUNDRY ROOM

Activities

Wash and dry clothes, location of furnace, storage for out of season clothes, iron.

Proximities

Kitchen, secondary exit, garage.

Requirements

Tables should be provided for folding and sorting

clothes, storage space should be provided for soaps and canned foods, storage is also needed for things like Christmas ornaments.

GARAGE

Activities

Park one car, storage, workshop.

Proximities

Sun porch, laundry room.

Requirements

One-car door, two car garage, the other half of the garage will be used for storage of bicycles, lawn furniture, and garden tools. The garage should be heated with a breezeway or sunporch between it and the house.

Design

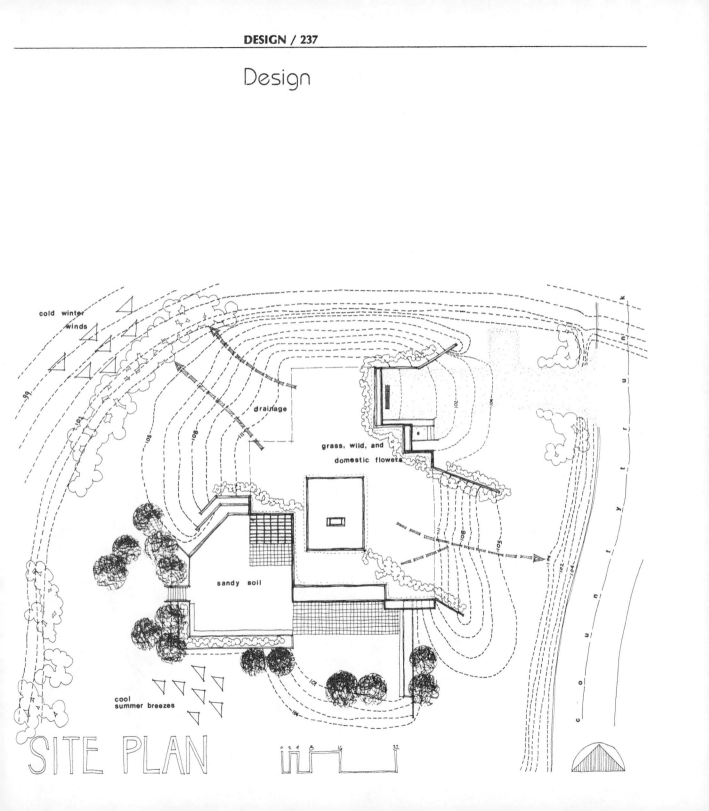

cold winter
winds

drainage

grass, wild, and
domestic flowers

sandy soil

cool
summer breezes

SITE PLAN

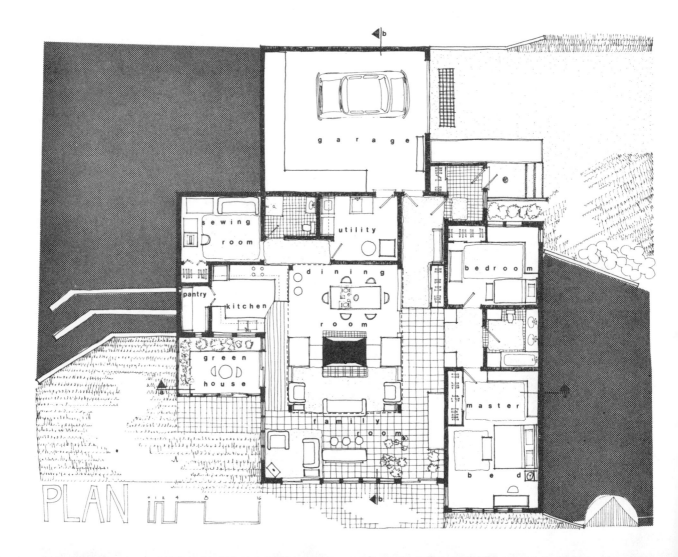

PLAN

SOUTH ELEVATION

EAST ELEVATION

SECTION aa

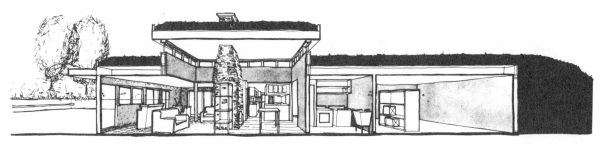

SECTION bb

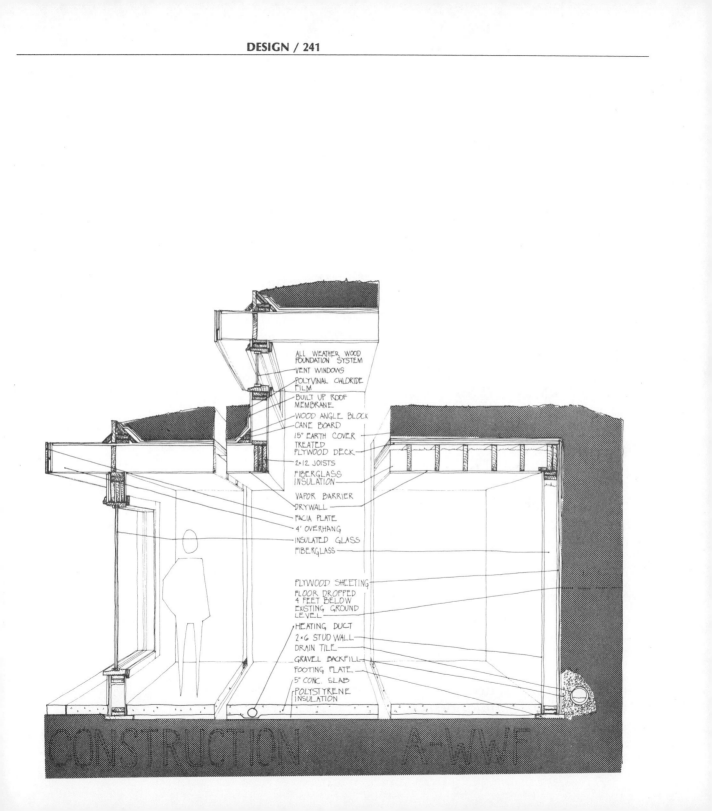

ALL WEATHER WOOD
FOUNDATION SYSTEM
VENT WINDOWS
POLYVINAL CHLORIDE
FILM
BUILT UP ROOF
MEMBRANE
WOOD ANGLE BLOCK
CANE BOARD
15" EARTH COVER
TREATED
PLYWOOD DECK
2×12 JOISTS
FIBERGLASS
INSULATION
VAPOR BARRIER
DRYWALL
FACIA PLATE
4' OVERHANG
INSULATED GLASS
FIBERGLASS

PLYWOOD SHEETING
FLOOR DROPPED
4 FEET BELOW
EXISTING GROUND
LEVEL
HEATING DUCT
2×6 STUD WALL
DRAIN TILE
GRAVEL BACKFILL
FOOTING PLATE
5" CONC. SLAB
POLYSTYRENE
INSULATION

CONSTRUCTION   A-WWF

BIBLIOGRAPHY

AIA Research Corporation. *Solar Dwelling Design Concepts.* U S Government Printing Office, 1976

American Institute of Timber Construction. *Timber Construction Manual.* John Wiley and Sons, Inc. 1974

Anderson, Bruce. *The Solar Home Book.* Cheshire Books, Harrisville, New Hampshire 1976

Beasley, R. P. *Erosion and Sediment Pollution.* Ames, Iowa, Iowa State University Press 1972

Bernard Rudofsky. *The Prodigious Builders.* Harcourt Brace Jovanovich 1977

Blendermann, Louis. *Design of Plumbing and Drainage Systems.* The Industrial Press 1959

Challender, John Handcock. *Time-Saver Standards for Architectural Design Data.* McGraw Hill Book Co. 1974

Ching, Francis. *Building Construction Illustrated.* Van Nostrand Reinhold Co. 1975

Eccli, Eugene. *Low-Cost Energy-Efficient Shelter.* Rodale Press 1976

Ellis Solar Construction. *Optimum Home Heating and Insulating Techniques.* *Farm Builder*, No. 8, May 1975

Flawn, Peter. *Environmental Geology.* Harper and Row 1970

Frisch, Karl. *Animal Architecture.* Excerpts and figures by Turid Holldobler from *Animal Architecture* by Karl von Frisch are reprinted by permission of Harcourt Brace Jovanovich, Inc.; copyright © 1974 by Karl von Frisch and Otto von Frisch; © 1974 by Harcourt Brace Jovanovich, Inc.; © 1974 by Turid Holldobler.

Here's the All Weather Wood Foundation System. The American Plywood Association 1973

Home Energy Digest and Wood Burning Quarterly. Fall 1978

Kinzey and Sharp. *Environmental Technology in Architecture.* Englewood Cliffs 1963

Lynch, Kevin. *Site Planning.* The Massachusetts Institute of Technology 1962

Ramsey and Sleeper. *Architectural Graphic Standards.* John Wiley and Sons, Inc. 1970

Malcolm Wells. *Underground Designs.* 1977

McGuinness and Stein. *Building Technology, Mechanical and Electrical Systems.* John Wiley and Sons, Inc. 1977

McGuinness and Stein. *Mechanical and Electrical Equipment for Buildings.* Fifth Edition. John Wiley and Sons, Inc. 1955

National Science Foundation Research Applications Directorate. *The Use of Earth Covered Buildings.* U.S. Government Printing Office 1976

Underground Space Center, University of Minnesota. *Earth Sheltered Housing Design.* University of Minnesota 1978

U.S. Government Printing Office. *Plants, People and Environmental Quality.*

White, Robert. *Landscape Development and Natural Ventilation.*